Periodicity and the s- and p-Block Elements

Periodicity and the s- and p-Block Elements

SECOND EDITION

Nicholas C. Norman

School of Chemistry, The University of Bristol

OXFORD
UNIVERSITY PRESS

UNIVERSITY PRESS

Great Clarendon Street, Oxford, OX2 6DP,
United Kingdom

Oxford University Press is a department of the University of Oxford.
It furthers the University's objective of excellence in research, scholarship,
and education by publishing worldwide. Oxford is a registered trade mark of
Oxford University Press in the UK and in certain other countries

Published in the United States of America by Oxford University Press
198 Madison Avenue, New York, NY 10016, United States of America

British Library Cataloguing in Publication Data
Data available

Library of Congress Control Number: 2020945106

ISBN 978-0-19-883534-9

Printed and bound by
CPI Group (UK) Ltd, Croydon, CR0 4YY

Preface to the First Edition

This book is a revised version of the Chemistry Primer entitled *Periodicity and the p-Block Elements*. In addition to having updated and substantially rewritten parts of the first book, the most obvious change is that certain aspects of s-block element chemistry are now discussed explicitly and so this group of elements are therefore included in the title. As in the first book, I have sought to provide an overview of some of the important trends (periodicity) found in the properties of the the s- and p-block elements and the compounds that they form, and to provide the reader with some simple 'rules of thumb' whereby they might appreciate and better understand these trends. In a book of this length, I have had to be selective and cannot possibly cover all aspects of periodicity (I have not included anything on solubility trends in s-block compounds for example), but I hope that those aspects which I have chosen will be both useful and informative. In reply to a certain 'reviewer' of the first book, I make no apologies for the non-mathematical treatment adopted in Chapter 1. If the student wants more, it is available in other texts listed in the bibliography, but I suspect that my qualitative and descriptive approach (in common with a great many other texts) will be perfectly adequate for most of the target audience.

As with the first book, I called upon many people to read the book in draft form, and for this I thank my colleage Guy Orpen together with Craig Rice, Edward Robins, Yvonne Lawson and Siân James who are all current members of my group.

Bristol   N.C.N.
February 1997

Preface to the Second Edition

This book is a second edition of the Oxford Chemistry Primer entitled *Periodicity and the s- and p-Block Elements* which was first published almost twenty-five years ago and has seen little revision throughout that time. Many parts of the book have been substantially rewritten and quite a lot of new material has also been added so the book is now a little larger. As with the first edition, I have sought to provide an overview of some of the important trends (periodicity) found in the properties of the the s- and p-block elements and the compounds that they form, and to provide the reader with some simple 'rules of thumb' whereby they might appreciate and better understand these trends. This includes a new chapter at the end which addresses some of the theories, models, and rules employed, particularly in terms of how we might know when they are appropriate and when not. In a book of this length I have had to be selective and cannot possibly cover all aspects of periodicity, but I hope that those topics which I have chosen will be both useful and informative.

As with the first edition, I have imposed upon some of my colleagues to offer suggestions and comments on the book in draft form. Particular thanks must go to Neil Allan and Paul Pringle who read the entire manuscript, and also to Fred Manby who read Chapter 1 and produced many of the graphs and figures used in that chapter. Simon Hall, David Fermin, and Chris Russell offered comments on those sections which cover the solid-state, electrochemistry, and oxidation states respectively. Since this text is aimed at students, I thought it essential that I ask an undergraduate student to read the entire draft, and would like to thank Elodie Heavens for so doing. I would also like to thank Hazel Sparkes for producing some of the structure figures. Any remaining errors, faults, or misconceptions are entirely down to me.

Bristol   N.C.N.
February 2020

Synopsis

1 Atomic structure and the form of the periodic table

This chapter deals with simple ideas concerning atomic structure and how these ideas can be used to understand the structure of the periodic table. The orbital model of the atom derived from a wave treatment of the electron is central to this approach.

2 Periodicity in the properties of s- and p-block atoms

This chapter considers, in the following order, effective nuclear charge, ionization energies, electron affinities, covalent and ionic radii (with some mention of van der Waals radii), electronegativity, orbital energies and promotion energies, and relativistic effects. Considerable emphasis is given to electronegativity, since this concept is used as an organizing principle in subsequent chapters.

3 Periodicity in the properties and structures of the elements

This chapter focuses on the structures of the s- and p-block elements paying particular attention to the trend from metallic to metalloid to non-metallic properties. It starts with a brief overview of the element structures and concludes with a section on general features which attempts to rationalize the observed trends, including some newer ideas on metallic bonding. This latter section includes a discussion of the effects of element electronegativity and solid-state band structure as a means of understanding the observed trends. A section on binding or atomization energies is also included.

4 General features of s- and p-block element compounds

This chapter starts with a general section on oxidation states and valence. The ensuing sections consider common oxidation states, including the inert pair effect and elements of the 4p row, followed by an outline of trends and stabilities of particular oxidation states. Element size and coordination numbers are then addressed, followed by an extensive section on bond energies dealing with trends in homonuclear and heteronuclear single bonds and multiple bonds. A comparison is made between trends in the s- and p-blocks compared to those observed in the d- and f-blocks. A final section deals with a general treatment of compound types in terms of the van Arkel–Ketelaar triangle or element triangle, and a section illustrating the so-called element tetrahedron which incorporates a discussion of van der Waals interactions between molecules in the solid-state.

5 Compounds of the s and p-block elements

This chapter deals with element halides, oxides, and hydrides, using specific examples to illustrate general trends in the properties of these compounds. Trends such as the change from ionic to polymeric to molecular are discussed in terms of element electronegativity differences, together with sections which explore ionic and solid-state structures in more detail. A final section looks at physical properties.

6 Acids and bases

Simple ideas on acids and bases such as the Lewis and Brønsted–Lowry definitions are presented, followed by a section on the acidity, basicity, and amphoterism of element oxides and hydroxides. A further section deals with the Lewis acidity of the heavier p-block elements discussed in terms of the vacant d-orbital and σ^*-orbital models. A concluding section deals with the concept of hard and soft acids and bases.

7 Structure

This chapter starts with a discussion of electron counting rules which leads on to a consideration of some general structural trends following an introduction to VSEPR theory. The trends in pyramidality and pyramidal inversion energy barriers for trivalent compounds of Group 15 are used as an example, together with the structures of heavier analogues of alkenes and alkynes and the observed structural trends. A final section deals with the Zintl concept for rationalizing many solid-state structures.

8 Theories and models: scope and limitations

This final chapter explores in a little more depth some ideas (some of them new) on bonding and the use of d orbitals, and concludes with a general discussion of the pros and cons of different models and approaches to understanding many of the ideas explored in this book.

Contents

Introduction

A large part of the fascination of inorganic chemistry lies in the almost limit-less variety and diversity of the compounds which the elements form amongst themselves and the myriad different properties they possess. Ionic solids, cova-lent molecules, metals and alloys, ceramics and polymers, all are examples of the types of materials encountered, and nowhere is this diversity more apparent than in the chemistry of the main group elements of the s- and p-blocks, the topic with which this book is concerned. Nevertheless, whilst variety and diver-sity are fascinating on the one hand, on the other it can all too easily be the case that main group chemistry is perceived as a vast collection of disparate facts any mastery of which can seem quite daunting.

The object of this book is to show that much of the huge body of knowledge which is s- and p-block chemistry can be understood and rationalized on the basis of a relatively few simple and straightforward principles. We start in Chapter 1 with an overview of ideas on atomic structure which lead us to an understanding of why the periodic table has the form that it does. It is no exaggeration to say that the periodic table is the single most important organizing principle in inor-ganic chemistry, and an understanding of its structure is therefore crucial if we are ever to understand the periodic relationships between the properties of the elements themselves and their compounds.

Chapter 2 deals with the periodic aspects of some of the properties of atoms either as isolated entities or as constituents of molecules. These include ioniza-tion energies and electronegativity, and we shall see that the latter is a power-ful organizing principle and can be viewed as a third dimension to the periodic table. The trends in all of the atomic properties we consider are seen to be closely related and thereby constitute a foundation upon which we can build an under-standing of the structures and properties of the elements themselves; this is the subject of Chapter 3.

Chapter 4 looks at some of the important general features associated with the compounds of the elements such as oxidation states, valence, bond ener-gies, and the physical nature of the compounds themselves, such as whether they are ionic, molecular covalent, polymeric, or metallic. This provides a basis to explore, in Chapter 5, the specific nature of selected classes of compound, such as the halides, oxides, and hydrides, and the factors which determine why these compounds have the characteristics they do. In all cases, the emphasis is on observed periodic trends and the reasons behind why such trends occur.

In Chapter 6 we look at the concept of acids and bases, and in Chapter 7 we consider some aspects of structure and bonding. In Chapter 8, which is new to this edition, we conclude with some further thoughts on bonding, but also address the issue of theories and approximate models, an appreciation of which is so essential to our understanding of the subject of this text.

A bibliography and a glossary are provided at the end along with some exercises at the close of most chapters which will help cement our understanding of the topics considered.

Atomic structure and the form of the periodic table

1.1 Introduction

An understanding of the structure or form of the periodic table is essential if we are ever to understand periodicity, and for this it is necessary to know something about atomic structure; in particular, how electrons behave in atoms. In this chapter we shall examine the topic of atomic structure in some detail and show how it is essential in explaining why the periodic table has the form that it does. More detailed treatments of this subject can be found in many of the inorganic chemistry textbooks listed in the bibliography though it will be evident from many of these texts that this topic is often dealt with in quite a mathematical way. Wave equations and the like are, of course, mathematical by their very nature, and a *full* understanding and appreciation of this subject cannot be had in the absence of a mathematical treatment. However, the basic ideas can be presented and understood using a descriptive and pictorial approach, and this is the one we shall adopt. Such an approach necessarily involves many *ad hoc* assumptions and statements which we will need to accept, but we will not suffer unduly because of this.

Initial ideas on atomic structure, which were developed in the early part of the twentieth century (after it was determined that atoms had any structure at all), consisted of a positively charged 'cloud' with embedded, negatively charged, particle-like electrons, the so-called 'plum-pudding' model proposed by Thompson. This was soon superseded by the Rutherford model in which a central, positively charged nucleus, containing most of the mass of the atom, was surrounded or orbited by the much lighter electrons. The resulting picture of the atom was classical in nature in that the atom was viewed as a miniature solar system in which the nucleus was analogous to the sun and the electrons to the planets. However, while this simple model was consistent with Rutherford's experimental results, it was at odds with certain aspects of classical physics (Maxwell's theory of electromagnetism) which predicted that the electrons in such an atom should spiral into the nucleus with concomitant emission of radiation, i.e. atoms ought to decay, and decay rather quickly!

We should recognize that the modern form of the periodic table was originally derived empirically (by Mendeleev and others) based on observed chemical similarities between certain groups of elements. A theoretical understanding of why the periodic table has the form it does came later (Section 1.6 provides a little more detail on this history).

A more detailed treatment of the subject of atomic structure can be found in Winter (2016). An alternative empirical approach has been described by Gillespie *et al.* (*J. Chem. Educ.*, 1996, **73**, 617) which emphasizes how the shell structure of electrons in atoms can be derived in a straightforward manner from photoelectron spectroscopy data rather than quantum mechanics.

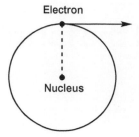

Fig. 1.1 A pictorial representation of the Bohr model of the atom.

An apparent solution to this problem came from contemporaneous observations in atomic spectroscopy which revealed that, for whatever reason, only certain orbits or energies were possible for the electrons. This formed the basis of the subsequent model of the atom proposed by Niels Bohr (Fig. 1.1), but whilst this model was very successful in explaining the observed spectroscopic properties of the hydrogen atom—the simplest of all atoms, containing only a single electron—it soon ran into problems with heavier atoms containing more than one electron. Furthermore, from a theoretical standpoint, the fact that only certain electron orbits (or energies) were possible or allowed had to be introduced on a purely *ad hoc* basis.

In 1926 an elegant solution to the problem was provided by Schrödinger who, building on the earlier work of de Broglie, proposed that electrons in atoms should be considered not as particles but as waves (for which there was experimental evidence), and that they could therefore be described by a suitable wave equation which would later bear his name. The important conceptual breakthrough was that the presence of only certain allowed electron energies is a direct and natural consequence of a wave treatment with no other assumptions necessary. We can see how this arises in a simple way by considering a one-dimensional model.

A suitable model for a one-dimensional wave-like system is a stringed musical instrument. When the string vibrates, a certain note is heard which has a fixed frequency, and it is therefore apparent that as a result of the string being constrained in some way (the constraint here is the length of the string), only a certain particular frequency of vibration is possible. Although it may not be immediately obvious, we can equate frequency (v) with energy ($E = hv$, where h is Planck's constant) which will be relevant in later discussions.

It is important to realize that even without changing the length or the tension of the string it is possible to sound other notes of higher frequency (the so-called higher harmonics), two of which are illustrated in Fig. 1.2, and, as with the frequency of the fundamental vibration, the frequencies of the higher harmonics are also determined by the constraints on the string. Thus, for a string with given constraints, only certain particular frequencies of vibration are possible. At this stage we can introduce the terms **node** (a point of zero amplitude) and **lobe** (displacement between nodes), and we should then recognize that the higher harmonics have extra nodes symmetrically disposed along the string (the two fixed ends of the string also have zero amplitude but by convention, nodes are places of sign change); the first harmonic has one extra node, the second harmonic has two, and so on (see Fig. 1.2). We shall take terms such as **amplitude**, **wavelength**, and **frequency** as understood.

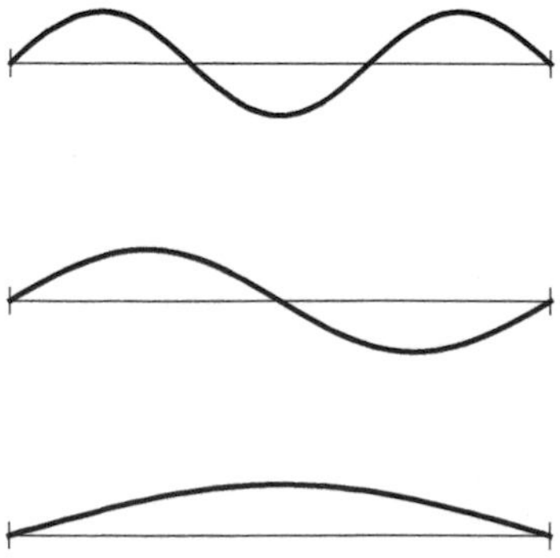

These waves which result from the vibrating string (illustrated in Fig. 1.2) may also be viewed as representations of solutions to a one-dimensional **wave equation** for which only certain solutions are possible. In other words, we may say that only certain **wavefunctions** (i.e. solutions to the wave equation) are allowed. Furthermore, these various solutions to the wave equation may be labelled with a number called a **quantum number** which can take only certain

Fig. 1.2 Vibrations of a string fixed at both ends. The fundamental vibration and the first two harmonics are shown. The horizontal lines indicate the position of the string at rest.

allowed values, each one of which corresponds to an allowed solution to the wave equation. It should therefore be apparent that this property of only certain allowed frequencies or energies (**quantization** as it is usually called) is a direct and natural consequence of a wave-like system which is constrained in some way, and that for a one-dimensional system there is *one* quantum number (we shall call it *n*) which can take various values (i.e. $n = 1, 2, 3$, etc.).

Returning now to atoms which are three-dimensional and, remembering that we are considering electrons as wave-like, there are three quantum numbers rather than one (we shall see later that other quantum numbers associated with the electron itself, i.e. independent of electrons in atoms, will also be necessary). It is these quantum numbers that we will use as labels for the wavefunctions or solutions to the aforementioned **Schrödinger wave equation** which is used to describe the behaviour or properties of electrons in atoms.

Before we look at three-dimensional systems in detail, let us review the important points again. The Schrödinger wave equation is used to describe the behaviour of electrons in atoms, i.e. we treat electrons as waves. Only certain solutions to the wave equation are possible, each corresponding to an allowed energy for an electron, and this quantization results directly from the system being constrained in some way. The constraints (often referred to as **boundary conditions**) for an electron in an atom are not as obvious as those for the strings discussed previously, but instead result from the attractive interaction between the positively charged nucleus and the negatively charged electron. Moreover, we obtain not one solution to the Schrödinger equation but many possible solutions, each described by a particular set of three quantum numbers. Each one of these solutions or wavefunctions (given the symbol ψ) describes a possible state of an electron in an atom and is referred to as an **orbital** (the **orbit** part is a hangover from more classical thinking about electron orbits), and each of these states has a distinct, discrete energy.

Let us now look at the solutions to the Schrödinger equation in more detail and see what they mean. In fact, the solutions or wavefunctions, ψ, refer to the **amplitude** of an electron wave more correctly termed a **probability amplitude**. Of more physical significance, however, is the square of this function, ψ^2 which is usually referred to as a **probability density** and is a measure of the electron density in a particular region of space. Thus ψ^2 at any particular point is the probability of finding the electron at that point (strictly speaking, the probability per unit volume of finding the electron close to the given point).

The terms 'quantum' and 'quantization' derive from **quantum mechanics**, which is the theory that describes the behaviour of electrons and all things very small. 'Probability' is fundamental to a quantum-mechanical understanding of electrons in atoms, but the probability of finding an electron at some point should not be thought of in terms of a particle model. We must remember that all of the treatment of electrons in atoms outlined here is based on their wave-like properties.

1.2 Wavefunctions for the hydrogen atom

Quantum numbers

We will start by considering the hydrogen atom, as it is the easiest system to look at since there is only one electron and the Schrödinger equation can be solved exactly; it can be used for other atoms with suitable modifications, as we shall see later. However, in order to allow us to represent much of what is to follow pictorially (something that is very useful to be able to do), it will be helpful to look at each wavefunction in two parts. Eqn. 1.1 shows how we can partition the wavefunction in this way, the two parts being the so-called **radial** and **angular** parts. We should also note at this point that because the hydrogen atom is a three-dimensional system, the wavefunction can be expressed using Cartesian coordinates x, y, z, though it is more convenient to use a spherical coordinate system defined by a distance, r, and two angles, θ and ϕ.

$$\psi = R_{n,l}(r)Y_{l,ml}(\theta, \phi) \tag{1.1}$$

The radial part, $R_{n,l}(r)$, depends only on the radial distance, r, between the nucleus and the electron and contains no information on direction or orientation. It depends on two quantum numbers, n and l, which will be defined shortly. The angular part, $Y_{l,ml}(\theta, \phi)$, depends on the direction or orientation, but not on distance, and is dependent on the quantum number l and on a third quantum number m_l, with the angles θ and ϕ defining an orientation within a spherical coordinate system.

We will not be concerned here with the details of how it is that these particular values for and relationships between the quantum numbers arise, why they are given these particular names, or even precisely why it is that three quantum numbers are required. It is the results in which we are interested. Some of the texts listed in the bibliography contain a more detailed account of this topic.

As stated previously, boundary conditions and the spherical three-dimensional nature of the atom give rise to three quantum numbers, and these, as we have just seen, are given the symbols n, l, and m_l. Each of the quantum numbers can take only certain allowed values, and a solution exists to the Schrödinger equation for particular allowed sets of these three numbers, with any one set of three defining a particular orbital or allowed energy state of an electron. The names and symbols for these quantum numbers, and the values which they can take, are given below.

(a) n is the **principal quantum number** and is associated with the radial part of the wavefunction. The energy of the orbital also depends on n (see later) but not on l and m_l. The number n can take integral values 1, 2, 3, 4 ... ∞, but we will be concerned only with the first few.

(b) l is the **azimuthal** or **angular momentum quantum number** and can take values 0, 1, 2, 3 ... $n - 1$, i.e. the possible values of l are dependent on n. The value of l influences both the radial and angular parts of the wavefunction and, in particular, determines the type or shape of the orbital which is usually given a letter designation.

$l = 0$ (s orbital); $l = 1$ (p orbital); $l = 2$ (d orbital); $l = 3$ (f orbital)

Orbitals are then labelled according to their value of n and the letter associated with l:

$$n = 1, l = 0\,(1s)$$
$$n = 2, l = 0\,(2s); n = 2, l = 1\,(2p)$$
$$n = 3, l = 0\,(3s); n = 3, l = 1\,(3p); n = 3, l = 2\,(3d)$$

Note that orbitals such as 1p and 2d are explicitly *not* allowed according to these rules.

(c) m_l is the **magnetic quantum number** and takes values $-l$, $-l+1$,... 0..., $l-1$, l, i.e. there are $2l+1$ values of m_l for a given value of l. Thus, for $l=0$, $m_l=0$, and so there is only one type of s orbital for any given value of n, i.e. one 1s, one 2s, etc. For $l=1$, $m_l=-1, 0, +1$, i.e. three types of p orbital for any given value of n except for $n=1$, i.e. three 2p, three 3p, etc. For $l=2$, $m_l=-2$, $-1, 0, +1, +2$, which means five types of d orbital. This quantum number specifies the orientation of the orbital within a defined coordinate system.

The particular numerical values of m_l are not especially important as far as this discussion is concerned, merely the total number of possible values allowed for any given value of l.

When we look at these orbitals or wavefunctions in more detail we will see that they contain nodes (just like strings), and there are a few simple rules concerning nodes that are worth noting at this point. The total number of nodes in any given orbital is equal to $n-1$. Some of these are in the radial part of the wavefunction, and some are in the angular part. The number of angular nodes is simply equal to the value of l, and so the number of radial nodes is therefore equal to $n-l-1$.

Now we will look at orbitals in more detail starting with the angular part of the wavefunction.

The angular part of the wavefunction

The angular part of the wavefunction reveals how the wavefunction (and hence electron density) varies as a function of the angles in a spherical coordinate system, and thus determines the shape of the orbital. It is dependent on the quantum number l, and, as we have seen previously, we can label the types of orbitals according to the value of this quantum number or more usually with the letters s, p, d, f, etc. Remember that the number of nodes in the angular part of the wavefunction (so-called angular nodes) for a given orbital is equal to l. Thus, s orbitals (1s, 2s, 3s, etc.) where $l=0$, have no angular nodes and, since there is no angular dependence of the wavefunction, the orbital is spherical in shape.

Why s, p, d, and f? Why not something like a, b, c, and d? They are old spectroscopic terms and actually stand for something: sharp, principal, diffuse, and fundamental.

For p orbitals, $l=1$ and there is one angular node. This node is a planar surface and divides the orbital into two lobes of opposite sign. There are three possible orientations for orbitals of this type (three values of m_l), and these lie along the axes x, y, and z (now in a Cartesian coordinate system), and are usually designated as p_x, p_y, and p_z. All p orbitals have this shape, and there are always three for any given value of n, i.e. three 2p, three 3p, etc.

For d orbitals, $l=2$ and there are now two nodes associated with these orbitals so that each therefore has four lobes. For a given value of n, there are five d orbitals (five values of m_l) which are given the labels d_{xy}, d_{xz}, d_{yz}, $d_{x^2-y^2}$ and d_{z^2}.

Pictures of the angular part of the wavefunction for a few selected orbitals are shown in Fig. 1.3, but remember that these are two-dimensional representations of three-dimensional orbitals; the angular part of the s orbital should be thought of as spherical. Note also that for the p_z orbital shown in Fig. 1.3(b) the xy plane is the nodal plane, and in the d_{xy} orbital shown in Fig. 1.3(c) the xz and yz planes are the nodal planes. The d_{z^2} orbital looks a little different and has two conical nodes (the two cones share an apex at the origin) but here we shall not be concerned with details of d orbitals; for more detail see Winter (2016).

As discussed in Section 1.1, we will generally be more interested in the square of the wavefunction, ψ^2, which tells us about the probability density or

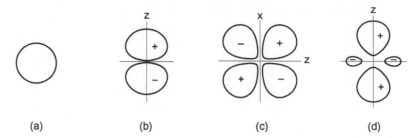

Fig. 1.3 Pictures of the angular part of the wavefunction for selected orbitals: (a) s; (b) p_z; (c) d_{xz}; (d) d_{z^2}. The lines represent the Cartesian axes x, y, and z. which are labelled where appropriate.

Fig. 1.4 Common representations of p orbitals.

The graphs of ψ and ψ^2 for p, d, and f orbitals, unlike those for s orbitals, are zero at the origin. A more complete set of graphs is given in Winter (2016) and in many of the standard texts listed in the bibliography. We will see later that there are some consequences which follow from the fundamentally different nature of s orbitals.

electron density. The graphs of ψ and ψ^2 for the angular part of the wavefunction do not look very different, but we should recognize that chemists generally represent orbitals in a somewhat inaccurate cartoon-like manner, albeit one which is very useful. Note also that in the graphs of ψ shown in Fig. 1.3, the signs of the wavefunction shown on either side of any nodes are represented as + and −. Think of the first harmonic shown in Fig. 1.2, which has one lobe above the baseline and one below; the one above we can represent as +, and the one below as −, which is what is meant by the sign of the wavefunction. An example of how orbitals as usually drawn is shown in Fig. 1.4 for a p orbital, one of which is shown with + and −, and the other with shading to indicate the relative sign of the wavefunction. It is important to realize that the + and − signs have *nothing* to do with charge. Furthermore, in graphs of ψ^2 everything is now positive; there cannot anywhere be a negative probability or electron density. A much fuller discussion of orbital types, including many excellent pictures, can be found in Winter (2016).

The radial part of the wavefunction

The radial part of the wavefunction tells us how the wavefunction varies with distance, r, from the nucleus, i.e. the effective size of the orbital, and we shall see that the square of this function will be particularly useful in understanding many aspects of the structure of the periodic table. Atomic orbitals depend on an exponential function e^{-Br}, where B is a constant and r is the distance from the nucleus. This means that ψ falls away exponentially with the value of r (and therefore never reaches zero). For a 1s orbital, a standard expression is $\psi = Ae^{-Br}$, and if we plot this function as a graph we obtain a curve shown in Fig. 1.5, from which we can see that there is a maximum value of ψ at $r = 0$ or at the nucleus. If we now plot a graph of ψ^2 vs r to determine the probability of finding an electron at a particular point, we find that this looks rather similar, as shown in Fig. 1.6.

However, we are not so much interested in the probability of finding an electron at a point along a line from the nucleus, but rather with the probability of finding the electron on a particular three-dimensional surface (strictly speaking, within the volume of a very thin shell). In the case of the 1s orbital being

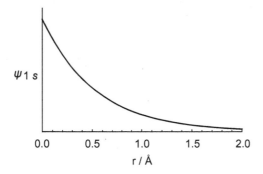

Fig. 1.5 Graph of ψ *vs* r for the 1s orbital.

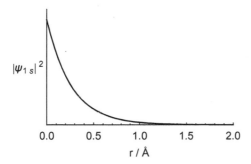

Fig. 1.6 Graph of ψ² *vs* r for the 1s orbital.

considered here, we can imagine a spherical surface (derived from the angular part) expanding from the nucleus, and it will be useful to know the probability of finding an electron anywhere on (or at some point on) this surface as a function of distance from the nucleus. The greater the distance from the nucleus, the more points there are on the sphere, and so the function we want will be proportional to the surface area of the sphere, or $4\pi r^2$. We can therefore plot a probability that an electron is on a surface at a certain distance r according to the function $4\pi r^2 \psi^2$ (more generally for all orbitals written as $R^2(r)$); a graph of this function for the 1s orbital is shown in Fig. 1.7(a).

This function $R^2(r)$ is called the **radial probability function** (or sometimes the **radial distribution function**). The inclusion of the $4\pi r^2$ term in $R^2(r)$ means that when $r = 0$, the radial probability function is zero, and so for all orbitals the function is zero at the nucleus. We see also that the graph has a maximum, i.e. there is a distance at which we are most likely to find the electron, r_o. For a 1s orbital for hydrogen, this is called the **Bohr radius** and is equal to 0.529 Å. This most probable distance is what is meant by the effective size of the orbital, but note that there is a small but finite probability that the electron can be found at large distances from the nucleus as a result of the exponential decay mentioned previously.

Graphs for the radial probability functions for hydrogen orbitals where $n = 2$ and 3 are shown in Figs. 1.7(b), 1.7(c), and Fig. 1.8 to illustrate some of the more important points. Note, for example, that some graphs have secondary maxima

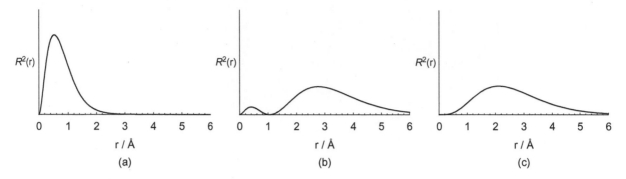

Fig. 1.7 Graphs of $R^2(r)$ vs r for the hydrogen atom: (a) the 1s orbital, (b) the 2s orbital, and (c) the 2p orbital, all drawn with the same vertical scale.

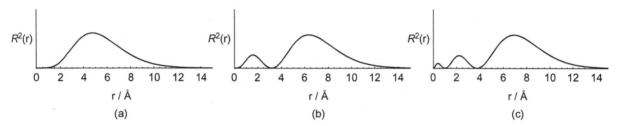

Fig. 1.8 Graphs of the radial probability functions for the hydrogen atom for: (a) the 3d orbital, (b) the 3p orbital, and (c) the 3s orbital drawn to scale. Note how the distance of the main maximum from the nucleus decreases with increasing l.

closer to the nucleus, and also that the major maximum in the graph of the 2s orbital is slightly further from the nucleus than is the maximum in the 2p orbital—a matter we shall remark upon later, although both are considerably further from the nucleus than that for the 1s orbital.

Some important points for all orbitals which derive from this treatment are as follows:

1. All functions tend to zero at large values of r.

2. Some functions have $R^2(r) = 0$ for certain values of r (other than for $r = 0$) which correspond to nodes (so-called radial nodes). Remember that the total number of nodes for any orbital is given by $n - 1$, and since the number of angular nodes is equal to l, the number of radial nodes is equal to $n - l - 1$.

3. All radial probability functions have a maximum at some value of r, i.e. r_o. If we compare 1s, 2s, and 3s we find that r_o increases (quite substantially) with n. Similarly, for 2p vs 3p vs 4p, i.e. the value of the principal quantum number n determines the size of the orbital. For a given value of n, the orbitals are of similar size, as shown in Figs. 1.7(b) and (c) and in Fig. 1.8, the main maxima being slightly closer to the nucleus the larger the value of l.

4. For orbitals with radial nodes, secondary maxima occur closer to the nucleus. This will be important when we look at polyelectronic atoms and the ordering of orbital energies.

It is important to remember that all of these Figures are graphs of mathematical functions and should not be taken as physical representations of how orbitals 'look'. If we try to represent both parts of the wavefunction together (i.e. the radial and angular parts) we obtain contour maps or shaded pictures, and in the contour plot of the 3p$_z$ orbital shown in Fig. 1.9 we see both a radial node and an angular node.

1.3 Energies of orbitals

In the case of the hydrogen atom, which contains only one electron, the energy of the electron can be calculated according to Eqn. 1.2, where Z is the nuclear charge (which is 1 in this case), R_H is the Rydberg constant, and n is the principal quantum number. Since Z and R_H are constants, we can say that the energy of the electron depends only upon n. Note that the energy is negative—more negative meaning lower in energy and therefore more stable and harder to remove.

$$E_n = -\left(Z^2 R_H\right)/n^2 \qquad (1.2)$$

The lowest energy or **ground state** of the electron in a hydrogen atom corresponds to the electron occupying the 1s orbital (where $n = 1$), so as well as talking about the energy of the electron we can also talk about the energy of the 1s orbital. The next-highest energy is where $n = 2$ such that the 2s orbital and the 2p orbitals (all three of them) have the same energy and are said to be **degenerate**. Similarly, the 3s, 3p, and 3d orbitals are all degenerate, as are 4s, 4p, 4d, and 4f, all of which can be represented diagrammatically in an energy-level diagram for the hydrogen atom which is shown in Fig. 1.10. Note that the separation in energy between the various levels is not constant, such that the levels become closer together as n increases which is a consequence of the energy scaling with $1/n^2$ rather than n.

Complete removal of the electron is defined as zero energy where $n = \infty$, and this corresponds to the **ionization energy** which is the difference between zero

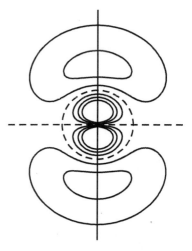

Fig. 1.9 A contour plot of the 3p$_z$ orbital (the vertical axis is the z axis). The angular node is represented by the horizontal dashed line (i.e. the xy plane) and the radial node by the dashed line circle. The sign of the wavefunction (not shown) changes every time a node is crossed.

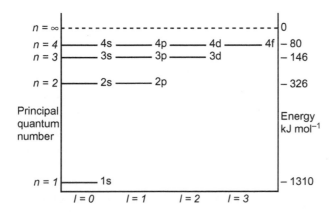

Fig. 1.10 An orbital energy-level diagram for the hydrogen atom.

and the energy of the 1s orbital for the hydrogen atom; for a one-electron system, the ionization energy corresponds exactly to minus the orbital energy. We can also note that the electron can occupy the higher-energy orbitals, in which case it is in an **excited state**. Transitions between energy levels result in emission or absorption of photons corresponding to what is observed in emission and absorption spectra.

1.4 Polyelectronic atoms

Having considered, in the preceding section, hydrogen with only a single electron, we are now ready to consider polyelectronic atoms, i.e. the rest of the periodic table. One immediate problem which we encounter is that the Schrödinger equation can only be solved exactly for the hydrogen atom, which has one electron (or other one-electron species such as He^+ and Li^{2+}). Solutions to the Schrödinger equation which describe polyelectronic atoms can only be solved approximately, although calculations involving various assumptions and approximations indicate that orbitals for other atoms are very similar to those found for hydrogen.

In particular, some important points are as follows:

1. The same quantum numbers are found as for hydrogen.

2. The same angular functions are found so that the orbitals have the same shapes and types, i.e. 1s, 2s, 2p, etc.

3. The radial functions are also similar, but are contracted to smaller radii (i.e. orbitals shrink due to increasing nuclear charge), and also, the energies are lower. Most importantly of all, orbital energies now depend on the azimuthal quantum number l as well as upon n for reasons which we shall address shortly.

An excellent chapter on orbitals in polyelectronic atoms, which gives more detail than is warranted here, can be found in Keeler and Wothers (2008).

Since we now have many electrons to deal with, and in order to understand the electronic structure of polyelectronic atoms and all that will follow from that, we must now look at how orbitals are filled by electrons. There are three rules or principles which we must consider.

The Pauli Exclusion Principle

There are actually two new quantum numbers to introduce at this point, s and m_s. The value of s is fundamental to the electron itself and always takes a value of ½, and so rarely features in any discussion of quantum numbers and atomic structure. m_s can take values of +½ and −½ as indicated in the main text. The relationship between s and m_s is the same as that between l and m_l; both are associated with angular momentum—orbital in the case of l, and spin in the case of s.

In order to understand the **Pauli Exclusion Principle** we must first introduce a fourth quantum number. This is labelled m_s and is associated with the electron itself and its **spin**. It can take the values +½ and −½, i.e. two distinct values. The Pauli Exclusion Principle in its simplest form states that **no two electrons in the same atom may have the same set of four quantum numbers**. Since each orbital is defined by three quantum numbers, n, l, and m_l, it follows that only two electrons can be associated with any particular orbital—one with $m_s = +½$, and the other with $m_s = −½$. Two such electrons are said to be **spin paired** or have **opposite spins**.

The *Aufbau* Principle

The ***Aufbau Principle*** states that the lowest-energy configuration is obtained when electrons go into the lowest-energy orbitals available. Thus, we build up (hence *aufbau*, from German) the electron configuration for an atom by adding electrons, two to each with spins paired, to the lowest-energy orbitals available. This is represented below for the first five elements in the periodic table which also shows the standard form of denoting the electron configuration of an atom. Note that 2p is higher in energy than 2s which we shall address in the next section.

$$H\,1s^1 \quad He\,1s^2 \quad Li\,1s^2\,2s^1 \quad Be\,1s^2\,2s^2 \quad B\,1s^2\,2s^2\,2p^1$$

Hund's First Rule

Hund's First Rule states that for a set of degenerate orbitals (orbitals with the same energy), electrons will start by going one into each with their spins **aligned** or **parallel**. This is shown below for the case of the filling of the 2p orbitals for carbon, nitrogen, and oxygen.

$$C\,1s^2\,2s^2\,2p_x^{\,1}\,2p_y^{\,1} \qquad N\,1s^2\,2s^2\,2p_x^{\,1}\,2p_y^{\,1}\,2p_z^{\,1} \qquad O\,1s^2\,2s^2\,2p_x^{\,2}\,2p_y^{\,1}\,2p_z^{\,1}$$

There are two reasons which are usually offered to account for this rule. Firstly, electrons go into different orbitals if possible since this places them further apart from each other (the $2p_x$, $2p_y$, and $2p_z$ orbitals are orthogonal) and thus minimizes the electrostatic repulsion which results from their having the same charge. Secondly, when two or more orbitals are singly occupied, as for carbon and nitrogen, spin parallel is lower in energy than spin paired (for quantum-mechanical reasons, spin parallel electrons repel less than spin paired electrons). A more complete account involves consideration of what is called **exchange energy**, a detailed description of which is beyond the scope of this text, though a very good account can be found in Keeler and Wothers (2008).

In concluding these three sub-sections, we can note that the maximum number of electrons in a shell is given by $2n^2$ (where n is the principal quantum number), and that the total number which can be accommodated in a sub-shell (i.e. s, p, d, or f) is equal to $2(2l+1)$.

The electronic structure of carbon has been written above as $2p_x^{\,1}\,2p_y^{\,1}$, but in terms of p orbital occupation this is entirely arbitrary, as it could just as well have been written as $2p_x^{\,1}\,2p_z^{\,1}$.

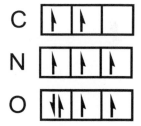

An alternative graphical representation of Hund's Rule is to use a set of boxes for the orbitals in which arrows denote electrons with the direction of the arrow indicating the electron spin.

1.5 Energies of electrons in polyelectronic atoms

As noted in the previous section, the energies of the electrons in non-hydrogen atoms, i.e. atoms with more than one electron, depend on both the quantum numbers n *and* l rather than just on n as is the case for hydrogen. We will now consider why this is so, and will see that an understanding of this fact is vital to understanding the structure of the periodic table.

Let us start by considering some first ionization energies, i.e. the energy required to remove the highest-energy or least tightly held electron. For hydrogen, the first ionization energy (1st IE) is 1313 kJ mol^{-1}. In comparison, the observed ionization energy for the isoelectronic helium ion (He$^+$) is found to be 5250 kJ mol^{-1}—both values consistent with what would be calculated using Eqn. 1.2. However, for the neutral helium atom (He) with two electrons, the observed 1st IE is found to be much less at 2372 kJ mol^{-1}. Clearly the two electrons in the 1s orbital of the He atom are at a higher energy (i.e. are more easily removed) than the single electron of He$^+$ which arises for a number of reasons. In He$^+$ there is a strong Coulombic attraction between the negatively charged electron being removed and the positively charged helium ion, and, with only a single electron in He$^+$, there are no electron–electron repulsions to consider. The electron–electron repulsion which does occur between the two s electrons in He, however, partly offsets the electron–nucleus attraction. We can therefore say that the electrons **shield** each other from the full attraction of the nucleus, and for He each electron feels a nuclear charge of only about +1.3 rather than the full +2 charge of the nucleus, i.e. partial but effective shielding of the electrons, by each other, from the positive charge of the nucleus.

For lithium, the observed 1st IE is 520 kJ mol^{-1}, which is quite a bit lower than for He. This suggests that there is a considerable shielding of the outermost electron in Li from the +3 nuclear charge, and the **effective nuclear charge** attracting this 2s electron of Li is only about +1.25 instead of +3. The outer 2s electron is therefore very effectively shielded from the nucleus by the two 1s electrons, and we can understand this in the following way.

If we look at the radial probability functions for the 1s and 2s orbitals superimposed on the same graph (see Fig. 1.11(a), which is a superposition of the graphs shown in Fig. 1.7(a) and (b)), we can see that most of the 2s orbital lies well outside the main part of the 1s electron density, i.e. the maximum for 2s is at a considerably greater distance from the nucleus than the maximum for 1s. The effect of this is that, in the case of Li, the two 1s electrons shield the 2s electron very effectively from the nuclear charge, though they provide much less shielding for each other. Moreover, the electron in the 2s orbital hardly shields the 1s electrons at all.

This leads us to consider the electrons in an atom in two groups. The **valence electrons** are those with the highest energies (normally those with the highest value of *n*) and are involved directly with chemistry and bonding, whilst the

We noted in Section 1.3 that for one-electron systems, ionization energy and orbital energy are the same thing. This is not the case for polyelectronic atoms which are discussed in Chapter 2.

Effective nuclear charge will be considered in more detail in Chapter 2.

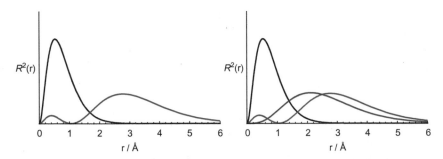

Fig. 1.11 (a) Graphs of radial probability functions for the 1s (black) and 2s (red) orbitals superimposed, and (b) graphs of the radial probability functions for the 1s (black), 2s (red), and 2p (blue) orbitals superimposed.

core electrons—those with lower values of n—are in completely filled orbital groups (shells) and act merely as a less-than-complete shield of the nuclear charge. The latter are too tightly bound to the nucleus to be directly involved in chemistry, i.e. their energy is too low, but they nevertheless play an important rôle as a result of their shielding influence on the valence electrons. In fact, the very existence of a periodic pattern in the properties and chemistry of the elements implies that it is only the valence electrons which are responsible for their chemical behaviour. It is therefore appropriate to think of the nucleus and the core electrons as providing an effective charge or potential in which the valence electrons move. This concept is very useful in simplifying many quantum-chemical calculations and is also fundamental to the depiction of orbitals in polyelectronic atoms.

In discussing the element lithium we placed the outermost electron in the 2s rather than the 2p orbital, and at this stage we need to consider why for Li, and all polyelectronic atoms, 2s is lower in energy than 2p. In other words, why do orbital energies now depend on n *and* l whereas for hydrogen they depend only upon n? The answer can be found by again considering the radial parts of the wavefunctions, and the particular importance of the small inner lobes or minor maxima seen in some of the radial probability functions.

For Li there are two 1s electrons and a third electron to be accommodated in an orbital of higher energy, so why does this go into 2s rather than the 2p? As we have seen, when considering the low ionization energy of Li, the two 1s electrons partly shield the third electron by getting between it and the nucleus, since the 1s electrons lie, on average, much closer to the nucleus (Fig. 1.11 (a)). However, a small but significant part of the 2s orbital lies even closer to the nucleus than the maximum in the 1s lobe. This is associated with the minor maximum in the radial probability function graph for 2s which results from the presence of a radial node. It is this contribution which lowers the energy of 2s relative to 2p, since the 2p orbital contains no radial nodes and hence no minor maximum close to the nucleus (Fig. 1.11 (b)). The electron in the 2s orbital can therefore be said to *penetrate* the core (i.e. 1s) better. The effect may seem to be small when looking at the graphs, especially when it is noted (as stated previously) that the major maximum for 2s is actually slightly further from the nucleus than the maximum for 2p, but it is sufficient and all-important in determining the ordering of the orbital energies. 2s therefore lies lower in energy than 2p, and it is the 2s orbital which is occupied first. This is the fundamental reason why the s-block comes before the p-block in the periodic table.

Similar arguments can be offered to explain why 3s penetrates more than 3p which, in turn, penetrates more than 3d, and therefore why these orbitals are filled in the order 3s first then 3p then 3d.

One point we should note at this stage is that we must be careful in using graphs of the radial probability functions for hydrogen atoms (which is what we have done in Fig. 1.11 and the associated discussion) to explain certain properties of polyelectronic atoms. As we shall see, graphs for a particular orbital can and do change markedly with respect to the distance of the maxima from the nucleus, depending on the nuclear charge (they contract with increasing charge, as noted previously) and on whether other orbitals are filled or not. In the case

In the s- and p-block, the statement that the valence electrons are those with the highest value of n is always correct. In the d- and f-block, as we shall see below, this is not always the case.

of Li and the 2s and 2p orbitals already described, the discussion based around the graphs in Fig. 1.11(b) is largely accurate because the radial extensions of the 2s and 2p orbitals are still similar (albeit contracted) even though 1s is filled; 1s, of course, is also contracted in Li compared to hydrogen. For the heavier or higher atomic number elements, it is no longer the case that ns and np orbitals have similar radial extensions, and it is found that as n increases, s orbitals have increasingly smaller radial extensions relative to p orbitals with the same value of n. This has important consequences for bonding as we shall see.

We are now in a position to start to appreciate why it is that the periodic table has the form it does. After H and He, the 2s orbital is filled (Li and Be) and then the three 2p orbitals, giving a total of six elements (B through Ne). This completes the filling of the orbitals of the second quantum shell (i.e. where $n = 2$) and, as mentioned previously, is why the p-block comes after the s-block in the periodic table, and also why the s-block has two columns and the p-block six columns. For the next element, Na, the outermost electron goes into the 3s orbital and the pattern of the previous shell repeats itself with the filling of 3s and 3p taking us as far as Ar. It is apparent from an inspection of the periodic table, however, that the outermost electron for the next element, potassium, resides not in one of the 3d orbitals, as might have been expected, but rather in the 4s orbital. So far, we have observed that orbital energies are dependent primarily on n with a secondary dependence on l. Clearly this is an oversimplification, since an orbital with $n = 4$ is being filled before all of the $n = 3$ orbitals have been filled.

We can go some way towards understanding this situation by considering the graphs in Fig. 1.12, which show the radial probability functions for the 4s and 3d orbitals and the extent to which they penetrate the 3s/3p core.

It is apparent that whilst 3d has its maximum closer to the nucleus than does 4s, the inner lobes of 4s (especially the innermost one) penetrate the core very effectively. Calculations reveal (though see later) that, with 3s and 3p filled, an electron in 4s is less efficiently shielded than an electron in 3d and therefore feels a slightly greater effective nuclear charge and so is slightly lower in energy. The outermost electrons for K and Ca therefore reside in 4s. Note, however, that once 4s has been filled it shields 3d very poorly, since much of the electron density in 4s lies outside the maximum of 3d. The 3d orbitals therefore fall sharply in energy as soon as 4s is

As noted previously, we must exercise some care in using the graphs shown in Fig. 1.12 which have been drawn for the hydrogen atom. For polyelectronic atoms the radial probability plots will change as a result of shielding and penetration effects. Thus, whilst for hydrogen the 4s orbital has a maximum much further from the nucleus than that of 3d, for elements like potassium these two orbitals are somewhat closer in size.

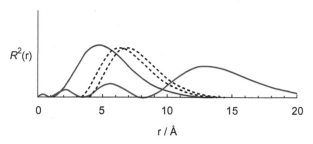

Fig. 1.12 Graphs of the radial probability functions for the 4s (red) and 3d (blue) orbitals superimposed on the 3s and 3p functions for which only the major maxima are shown (black dashed lines).

filled, and these five orbitals are the next to be occupied, which gives rise to the ten elements of the first transition series, Sc through Zn (and is why the d-block has ten columns). The next lowest-energy orbitals available are the three 4p orbitals, and these are filled for the elements Ga through Kr. This account therefore provides us with an understanding of why the d-block comes between the s- and p-blocks in the periodic table, and the rest of the periodic table can now be built up and understood in much the same way. The situation becomes a little more complicated with the later transition elements and the lanthanides and actinides (strictly, lanthanoids and actinoids) because there are many orbitals that are very close in energy, but we will not be concerned here with this part of the periodic table.

An important point which follows from this discussion is that the energies of orbitals (or, to be more precise, the order in which they are filled) are very dependent, as a result of shielding and penetration effects, on whether or not other orbitals are filled or unfilled (we pointed this out previously). A very useful graph of the variation in the orbital energies for the elements is shown in Fig. 1.13, and we shall refer to this again in subsequent chapters.

From Fig. 1.13 we can see that for hydrogen we have the special simple case that E depends only on n and so 2s and 2p are degenerate, as are 3s, 3p, and 3d. For all other atoms, the s, p, and d orbitals for any given n are *not* degenerate, due to differences in penetration and shielding effects. All s orbitals penetrate the inner shells well, so they feel a higher effective nuclear charge than corresponding p and d orbitals; they all drop steadily in energy as the nuclear charge increases (black lines). All p orbitals also penetrate the inner shells well except that they are well shielded from the nucleus by the 1s electrons so they drop steadily after staying nearly level for the first few elements. Thus, p orbitals (blue lines) are a little above s orbitals. 1s, 2s, 2p, 3s, and 3p maintain the same order for all elements.

In Fig. 1.13, the 3d orbital is drawn at higher energy than the 4s for elements with an atomic number of around 20–30. Whether or not the 3d orbital ever does rise above 4s in energy is, in fact, a matter of some debate, but the graph and associated discussion are adequate for our purposes. In any event, we must be careful to distinguish between statements about orbital energies and about the order in which orbitals are filled.

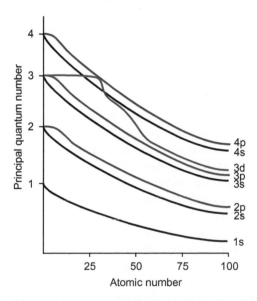

Fig. 1.13 Approximate orbital energies up to 4p for the elements up to $Z = 100$. s orbitals are shown in black, p orbitals in blue, and the 3d orbital in red.

Returning to 3d orbitals, note that once they are filled, they drop quickly in energy (red line) and approach 3s and 3p since they are very poorly shielded by the next orbitals being filled, i.e. 4p. Thus, for Ga and beyond, 3d are now much lower in energy than 4s and act as core rather than valence electrons. Therefore, the electron configuration of Ga is $[Ar]3d^{10}4s^24p^1$, and it resembles Al, $[Ne]3s^23p^1$, in its chemistry. The d orbitals do play a part in Ga chemistry but only a minor and indirect one which we shall look at in more detail in later chapters.

Two final points are important regarding Fig. 1.13.

1. Note that for core orbitals the individual s, p, and d orbitals with the same value of n come close together in energy with increasing atomic number; i.e. they become more hydrogen-like.

2. All of the previous discussion is for neutral atoms only; for ions, the ordering can and does change.

The second point above regarding neutral atoms *vs* ions requires a little more explanation. As we have seen, K has the valence electronic configuration $4s^1$, and Sc, the first element of the transition series, is $4s^23d^1$ (for gas phase atoms). It is generally true for the first-row transition elements that the neutral atoms have a ground state electronic configuration of $4s^23d^n$ except for Cr, which is $4s^13d^5$, and Cu, which is $4s^13d^{10}$. The main reason usually given to account for the specific electronic configurations of Cr and Cu is that there is a certain extra stability associated with half-filled and filled d shells (similar for p and f shells as well), and this should be sufficient to remind us that the lowest-energy or ground state configuration of an atom is not just based on orbital energies alone but on the total energy of the system, which will include inter-electron repulsions (electron correlation) etc. We should therefore expect that in some cases higher-energy orbitals will start to be filled before lower-energy orbitals are completely filled if they are close in energy and therefore not strictly according to the Aufbau Principle; this is strongly reminiscent of high spin *vs* low spin situations in transition metal complex chemistry.

However, in discussing the cations of the first-row transition elements, it is found that the 3d orbitals are lower in energy than the 4s orbital (i.e. are filled first) and the valence electron configuration of the metal +2 and +3 ions are always given as $3d^n$. This is often confusing in that it seems that whilst electrons are added to 4s and then 3d for the atoms, they are removed in the reverse order to form the ions. We can resolve this apparent difficulty by recognizing that the relative ordering of the orbital energies can be, and in this case is, dependent on the charge of the atom or ion. Thus, whilst 4s is lower than 3d for the neutral atoms, the reverse is true for the +2 and +3 ions (in the gas phase). The reason for 3d being lower for the ions is that the effective nuclear charge felt by the outer electrons is now much larger with the result that all orbitals become contracted. Penetration effects therefore become less important, and orbitals become more hydrogen-like such that energies become more dependent on just n and less dependent on l exactly as we see in Fig. 1.13.

Explanations for the ground state electronic structures of Cr and Cu involving the stability of half-filled and filled shells are more correctly argued in terms of exchange energies as we saw when discussing Hund's First Rule.

Some calculations indicate that 3d is always lower in energy than 4s, and the reason why 4s is populated before 3d is fully filled (in neutral atoms) is because inter-electron repulsions are significantly less between electrons in the 4s orbital due to its greater size. In cations the 3d orbitals are sufficiently stabilized relative to 4s that any advantage of reduced inter-electron repulsions in 4s is outweighed by the lower energy of 3d.

Before leaving this topic we should note that much of the previous discussion is captured in the so-called **Madelung Rule** which states that orbitals with a lower value of $n + l$ are filled first. This is often illustrated graphically in the type of chart shown in Fig. 1.14, but we should recognize that the value of this 'rule' is often questioned in light of some of the exceptions noted previously. Eugen Schwarz and co-workers (*J. Chem. Ed.*, 2010, **87**, 435 & 444) have addressed this topic in some detail and exposed some important shibboleths.

1.6 The structure of the periodic table

In concluding this chapter we are now able to appreciate some of the key features of the periodic table, particularly in terms of the relative positions of the s-, p-, and d-blocks. In Chapter 2 we will go on to look at the periodic trends in various atomic properties, but before doing so it will be useful to say a little more about the details of the structure of the periodic table, if for no other reason than that there are still discussions and disagreements about where some elements should be placed.

This is not the place to review the fascinating history of the periodic table (celebrated in 2019 by UNESCO as the International Year of the Periodic Table to mark the 150th anniversary of Dimitri Mendeleev's first publication on the topic), but we will say a little more about its structure.

The first point to reiterate is that the periodic table was initially constructed based on a consideration of atomic weight (nowadays referred to as relative atomic mass) and periodic trends in the chemical behaviour of the elements (chlorine similar to bromine similar to iodine, for example). It was therefore arrived at empirically, and only later were the underlying reasons for its structure understood with the development of quantum theory in the early twentieth century; many of the details have been outlined previously. Thus, we can appreciate and understand fundamental features such as the so-called s-, p-, d-, and f-blocks having 2, 6, 10, and 14 groups or columns respectively, and the fact that this arises from s orbitals being able to accommodate two electrons, p orbitals, a total of six for a given value of n, d orbitals ten, etc. We can also see why it is that most periodic tables place the d-block between the s- and p-blocks. If it is the ground state electronic structure of neutral gas phase atoms that is to be our primary criterion (though not everyone agrees that it should be), element 19 (K) with an electronic configuration of $[Ar]4s^1$ clearly should reside in the s-block, and element 21 (Sc) with a configuration $[Ar]4s^2 3d^1$ starts the d-block immediately after element 20 (Ca, $[Ar]4s^2$); the p-block then follows with element 31 (Ga, $[Ar]4s^2 3d^{10} 4p^1$) after the first d-row concludes with element 30 (Zn, $[Ar]4s^2 3d^{10}$). The fact that Cr and Cu are, respectively, $s^1 d^5$ and $s^1 d^{10}$ rather than $s^2 d^4$ and $s^2 d^9$ in their ground state makes little difference to this argument and merely reflects how close in energy 4s and 3d are (and the effects of electron-electron repulsion). Remember that we have seen earlier in this chapter why 4s is filled before 3d in neutral atoms.

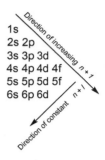

Fig. 1.14 The order of orbital filling according to the Madelung Rule. Orbitals are filled in the order shown by the arrow pointing to the lower left before starting on orbitals with a higher value of $n + l$.

A complete periodic table with the group numbering system is presented on the inside of the back cover. For an excellent history and general discussion, see *The Periodic Table* by Eric Scerri, Oxford University Press (2007).

If potassium were $[Ar]3d^1$ rather than $[Ar]4s^1$ (i.e. if 3d were filled before 4s), the d-block would come after the p-block.

When it comes to the f-block, we can advance similar arguments as to why this comes after the s-block and before the d-block along the same lines as we did for why the d-block comes after the s-block, but matters are more controversial here. In some, particularly older, periodic tables, the Group 3 elements are shown as Sc, Y, La, and Ac and the f-element rows start and end with Ce and Lu (4f) and with Th and Lr (5f). In some other versions of the periodic table, however, the Group 3 elements are listed as Sc, Y, Lu, and Lr, with the 4f and 5f rows changed accordingly. The IUPAC approved version of the periodic table avoids this issue by having two gaps under Y (there is no controversy over where Sc and Y should be) with the lanthanides and actinides comprising rows of fifteen elements, La to Lu and Ac to Lr respectively. The reason for differing views on the placement of some elements is largely about whether the electronic structure of the atom or its chemical/physical properties should have primacy in determining where an element is placed. Since this is not a text on d- or f-block chemistry, we will not dwell on the details which support either viewpoint, but it is important to note that contrary to what some might assume, there is no universally accepted version of the periodic table—no 'correct' version. Indeed, a quick survey of the literature will show that there are all sorts of different forms.

Some interesting versions of the periodic table can be found at: https://www.webelements.com

What has been discussed above centres on issues in the d- and f-blocks, and might seem of little consequence with regard to the s- and p-blocks which are the focus of this text. However, whilst the s- and p-blocks are largely uncontroversial, the positions of hydrogen and helium are a matter of disputation. Most would accept that hydrogen does indeed belong in Group 1 since it has a single valence s-electron just like all the other Group 1 elements (although in older tables, one sees it in a place on its own or sometimes even at the top of the halogens). Helium, however, is more problematic. What is it doing in the p-block? It does not have any p-electrons! It does have a closed shell, however, and its chemical properties are those of a noble gas just like all the other Group 18 elements. Nevertheless, with an s^2 pair, why not have it in Group 2 above beryllium? This is precisely where some periodic tables do place it!

To conclude, there is no universally accepted version of the periodic table, there probably is not going to be, and furthermore we should not expect that there should be. The positioning of most elements is uncontroversial, certainly in terms of their relationship to their neighbours, but arguments will persist about how the blocks should be organized and about some of the elements at the start or end of some of these blocks. We shall now move on to examine some important atomic properties.

2 Periodicity in the properties of s- and p-block atoms

In this chapter we will look at a range of properties of isolated atoms of the s- and p-block elements which will introduce us to the ideas of **periodicity**, and which will be useful later when we consider both the elements themselves and the compounds they form.

2.1 Effective nuclear charge

In Chapter 1 we introduced the concept of **effective nuclear charge** when considering the energies of the valence electrons, and from this it was clear that the outermost electrons in an atom feel a nuclear charge which is substantially less than the actual nuclear charge due to the shielding effects of other electrons. We shall see that this idea of effective nuclear charge is useful in understanding many aspects of periodicity, but in order to utilize and appreciate the concept fully, a quantitative scale (at least an approximate one) is desirable whereby we can look at values and trends in values for the valence electrons of the elements in which we are interested.

The first point we can note is that an effective nuclear charge, often given the symbol Z^* (but sometimes Z_{eff}), can be determined if we know the relevant ionization energy or orbital energy according to Eqn. 2.1. This is essentially the same as Eqn. 1.2 in Chapter 1 when we were considering hydrogen (the difference between the two equations being Z vs Z^*).

Ionization energies and orbital energies are not exactly the same for polyelectronic atoms (they are usually close) for reasons we shall see later.

$$E_n = -\left(Z^{*2} R_H\right)/n^2 \qquad (2.1)$$

There are also a number of schemes which have been devised to estimate values of Z^*, one of the simplest and most widely used (because of its simplicity) being a set of rules devised by Slater and known, not surprisingly, as **Slater's rules**. The object of Slater's rules is to estimate, for a particular electron in an atom (or ion), the strength of the shielding effect of the other electrons present, and from this to calculate a shielding constant, S. This value can then be used to calculate the effective nuclear charge, Z^*, according to Eqn. 2.2, where Z is the actual nuclear charge:

$$Z^* = Z - S \qquad (2.2)$$

To calculate the shielding constant, S, for a particular ns or np electron (Slater's rules make no distinction between them), the following rules apply:

1. Electrons with a higher n contribute zero, i.e. no shielding.
2. Electrons with the same value of n contribute 0.35, i.e. not very good shielding.
3. Electrons with a value of n one less than our chosen electron contribute 0.85, i.e. rather better shielding.
4. Electrons with lower values of n contribute 1.00, i.e. complete shielding.

For d and f electrons, the rules are slightly different, but we will not be concerned with the d- and f-block elements here.

A couple of examples will illustrate how the rules work. For lithium, the electron configuration is $1s^2 2s^1$. Thus, for the 2s electron, the shielding constant is $2 \times 0.85 = 1.70$. Hence Z^* for the Li 2s electron $= Z - S = 3 - 1.70 = 1.30$. In the case of nitrogen, the electron configuration is $1s^2 2s^2 2p^3$. Therefore, S for one of the 2p electrons is 4×0.35 (i.e. the two remaining 2p electrons and the 2s pair) $= 1.40$, plus 2×0.85 for the 1s pair, giving a total of 3.10. Hence $Z^* = 7 - 3.10 = 3.90$.

Values of Z^* for the s- and p-block elements calculated using Slater's rules are given in Table 2.1. A particularly important trend to note is that the effective nuclear charge increases quite substantially on crossing a period from left to right, the importance of which we shall see later. Values increase to a much smaller extent on descending a group and, because of how the values are calculated according to the rules, become constant lower down.

In considering the values in Table 2.1, it should be stressed that Slater's rules are very approximate and do not take into account such factors as the difference in the extent to which electrons in s and p orbitals penetrate the core electron density.

Other rules, such as those devised by Clementi and Raimondi, deal explicitly with effects ignored by Slater's rules but at the expense of simplicity. Thus, while Clementi and Raimondi Z^* values also show a marked increase from left to right, rather than values becoming constant down a group, they continue to increase (see Keeler and Wothers (2008) for more detail).

One might ask why, if Z^* values can be calculated according to Eqn. (2.1), scales such as those of Slater or Clementi and Raimondi are necessary. It is certainly the case that tables of Z^* values based on experimental ionization energies or calculated orbital energies can be produced, but approximate methods are still useful in understanding and rationalizing basic trends.

Table 2.1 Calculated values of Z^* from Slater's rules for the valence ns or np electrons of the s- and p-block elements

Li	Be		B	C	N	O	F	Ne
1.30	1.95		2.60	3.25	3.90	4.55	5.20	5.85
Na	Mg		Al	Si	P	S	Cl	Ar
2.20	2.85		3.50	4.15	4.80	5.45	6.10	6.75
K	Ca		Ga	Ge	As	Se	Br	Kr
2.20	2.85		5.00	5.65	6.30	6.95	7.60	8.25
Rb	Sr		In	Sn	Sb	Te	I	Xe
2.20	2.85		5.00	5.65	6.30	6.95	7.60	8.25
Cs	Ba		Tl	Pb	Bi	Po	At	Rn
2.20	2.85		5.00	5.65	6.30	6.95	7.60	8.25

2.2 Ionization energies

The ***ionization energy*** (sometimes called ionization potential or electron bind-ing energy) is the energy required to completely remove an electron from an atom in the gas phase. For example, Eqn. 2.3 illustrates the first ionization energy of a neutral atom, E. Ionization energies for all neutral atoms are endothermic, i.e. they require energy.

$$E(g) \; \rightarrow \; E^+(g) + e^- \tag{2.3}$$

The second ionization energy will be the energy required to remove an electron from the singly charged cation E^+ according to Eqn. 2.4. In all cases, the second ionization energy will be greater than the first, since it is more difficult to remove an electron from a positively charged ion than from a neutral atom, as we saw for He *vs* He^+ in Chapter 1.

$$E^+(g) \; \rightarrow \; E^{2+}(g) + e^- \tag{2.4}$$

Let us look first at the overall trends in first ionization energies of the s- and p-block elements, which are presented in Table 2.2. It is apparent from Table 2.2, and also from the graph shown in Fig. 2.1, that there is a general tendency for ion-ization energies to increase as we cross the periodic table from left to right and to decrease as a group is descended. Moreover, a sharp drop occurs when we start a new row (i.e. go to the next higher value of *n*). The trend across a period is to be expected on the basis of the effective nuclear charges discussed in Section 2.1 since, as Z^* increases across the period, the electrons will be more tightly held and therefore require more energy to remove. However, it is clear from Fig. 2.1 that the trend in ionization energies is not linear as might be anticipated from Slater's rules alone (see below). The sharp drop which occurs on starting a new

Table 2.2 First ionization energies (kJ mol^{-1}) for the s- and p-block elements

H 1312								He 2372
Li 513	Be 899		B 801	C 1086	N 1402	O 1314	F 1681	Ne 2081
Na 496	Mg 738		Al 578	Si 786	P 1012	S 1000	Cl 1251	Ar 1520
K 419	Ca 590		Ga 579	Ge 762	As 947	Se 941	Br 1140	Kr 1351
Rb 403	Sr 549		In 558	Sn 709	Sb 834	Te 869	I 1008	Xe 1170
Cs 376	Ba 503		Tl 589	Pb 715	Bi 703	Po 812	At 930	Rn 1037

Values in Table 2.2 are taken from Emsley (1989). Note that all values are positive, i.e. ionization requires energy (is endothermic).

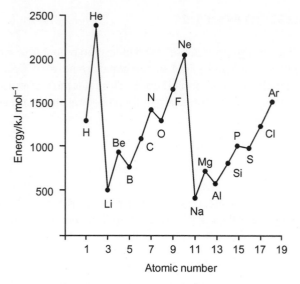

Fig. 2.1 First ionization energies for the elements hydrogen to argon.

Within a group, the importance of the dependence of the ionization on the value of *n* and hence distance from the nucleus is a consequence of Coulomb's Law, which states that the attraction between two opposite charges decreases in proportion to $1/r^2$.

row (e.g. Ne to Na) is also to be expected on the basis of Z^*, but the trend down a group is perhaps unexpected since values for Z^* increase rather than decrease down a group, before becoming constant (or continuing to increase if calculated using Clementi and Raimondi rules). In fact, the most important factor in a group is not so much the effective nuclear charge that the electron feels, but rather its distance from the nucleus. Thus, we should recall from Chapter 1 that the maximum in the radial probability function graph for an orbital is very dependent on the value of the principal quantum number *n*, the average distance from the nucleus increasing substantially with increasing *n*. Thus, even though the effective nuclear charge is increasing down a group, this is outweighed by the fact that the outer electron is increasingly further away from the nucleus. It is this latter factor which dominates and is therefore why electrons are easier to remove as groups are descended.

If we now look at Fig. 2.1 in more detail, we can see that whilst the ionization energies for Li to Ne generally increase as expected on the basis of Z^* values, there are kinks in the graph at boron and oxygen, and we shall consider each of these in turn.

In the case of boron with a lower first ionization energy than beryllium, the standard explanation is that the single 2p electron in boron is quite effectively shielded by the 2s pair and is therefore more easily removed (see Fig. 1.11(b) and note the minor maximum close to the nucleus for the 2s orbital)—an aspect of shielding not taken into account in Slater's rules, so this discrepancy is not unexpected.

In the case of oxygen, which clearly has a lower ionization energy than nitrogen, the explanation is a little more complicated. The reason often advanced is that whereas nitrogen has one electron in each of the three p orbitals, electrons have to pair up for oxygen and one of the p orbitals becomes doubly occupied.

Two electrons in one orbital are confined to the same region of space and thus experience a greater coulombic repulsion and it is this which accounts for the lower than expected first ionization energy of oxygen. Furthermore, ionization of oxygen results in a stable half-filled p shell, such a shell being disrupted when nitrogen is ionized. A much more satisfactory explanation, although one that is a little less intuitively obvious, involves a consideration of electron exchange energies first mentioned in Chapter 1—something which derives from a quantum-mechanical treatment of electrons and which has no classical analogue. A detailed treatment is beyond the scope of this text (see Keeler and Wothers (2008) for a very readable explanation), but after an analysis of exchange energies it is clear that it is boron and oxygen that have the expected ionization energies and that carbon, nitrogen, fluorine, and neon all have values higher than expected due to an unfavourable loss of exchange energy which occurs on ionization.

Exchange energies are associated with the number of electrons which have parallel spins (the more, the better) and is the basis for Hund's First Rule described in Chapter 1.

After neon there is a sharp drop on moving to sodium, as expected on the basis of the previous discussion, since Z^* values for Ne and Na are estimated to be 5.85 and 2.20 respectively, and also we have moved from the second quantum shell to the third and therefore the outermost electron in sodium is now much further from the nucleus. The subsequent trend from sodium to argon is very similar to that seen for lithium to neon, and for very much the same reasons.

Let us now look in more detail at trends within a group which we can illustrate by considering Groups 13 and 18. In Group 18 (and in Groups 1 and 2) there is a regular decrease in the first ionization energy as the group is descended due primarily to the increase in the principal quantum number associated with the outermost electrons, and the resulting increase in the distance of this electron from the nucleus. As mentioned previously, this factor is more important than the increase in Z^* as the group is descended.

If we look at Group 13, however, there at first seems to be little regularity in the ionization energies. Aluminium is lower than boron as we would expect, but the value for gallium is actually slightly greater than that for aluminium. The obvious difference between B vs Al and Al vs Ga is that Ga is preceded by the first row of the d-block elements, and it is to this that we should look for an explanation. Thus, in the absence of the 3d row, gallium would come after calcium and we would expect a regular trend from B to Al to Ga. With the inclusion of the elements Sc through Zn, however, we have added ten units of positive charge to the nucleus, but the corresponding ten electrons are in 3d orbitals which do not provide very effective shielding for the 4p electrons since 4p, with its minor maxima close to the nucleus (like 4s), penetrates 3d quite well. As a result, the effective nuclear charge felt by the 4p electron in gallium is larger than it would be in the absence of the 3d row, thereby making it harder to remove. Slater's rules as they are given in Table 2.1 are not particularly helpful in this regard because, although Z^* for Ga is larger than that for Al, the same is true for Al vs B. Nevertheless, we can use these rules as a guide since, in the absence of the 3d row (i.e. if Ga were [Ar]$4s^24p^1$), gallium would have a Z^* of 3.50 rather than the significantly larger value of 5.00 given in Table 2.1 which takes account of the effect of the 3d electrons. Clearly, the increase in Z^* more than offsets the increase in n in determining the ionization energy in the case of Ga vs Al.

Note that the ionization energies for the series of elements B, Al, Sc, Y, and La, i.e. Group 13 then Group 3 (although note the comments on the Group 3 elements in Section 1.6) show a much more regular decrease since there are no disparate shielding effects resulting from different core electron configurations.

Although Groups 1 and 2 appear to show a rather smooth trend in decreasing ionization energies, it is worth noting that Fr (400 kJ mol^{-1}) and Ra (509 kJ mol^{-1}) each have higher first ionization energies than Cs (376 kJ mol^{-1}) and Ba (503 kJ mol^{-1}) respectively which continues to reflect the presence of the 4f row as well as the increasing importance of relativistic effects.

On moving further down the group, we see that the ionization energy for indium is less than gallium—the general trend we would expect—but at thallium a substantial increase occurs such that the first ionization energy of thallium is larger than for all the elements in this group except boron! Two factors are important here. Firstly, the inclusion of the 4f row will lead to an increase in the effective nuclear charge felt by a thallium 6p electron for exactly the same reasons as those given above for the effect of the 3d row on a 4p electron. Secondly, so-called relativistic effects, to be described later, result in a lowering of the energy of the 6p orbital.

The increasing effective nuclear charge as a result of d and f rows (described above) plus relativistic effects (described later) have important consequences for the chemistry of the 4p and 6p elements as we shall see. Although the effects are most pronounced in terms of ionization energies for the elements of Group 13, we can also see the effects present in Groups 14–16. Thus, in Group 14, the ionization energy of Ge is very similar to that of Si, and Pb has a higher first ionization energy than Sn. In Groups 15 and 16, the similarity in ionization energies between the 3p and 4p and between the 5p and 6p elements is apparent.

We will return to ionization energies later when discussing the chemistry of the elements where trends in the sums of particular ionization energies will be important.

2.3 Electron affinities

By convention, electron affinities are given a *positive* value when the addition of an electron is exothermic. This somewhat confusing situation derives from a view that an exothermic electron attachment represents a 'positive affinity'. An alternative way of looking at electron affinities which avoids any confusion over sign is to regard them as *zeroth ionization potentials* defined according to the reverse of Eqn. 2.5, i.e. $E^-(g) \rightarrow E(g) + e^-$.

The **_electron affinity_** of an atom is defined as the energy change which occurs when an electron is added to an atom (or ion) according to Eqn. 2.5; it is the reverse of ionization. Experimental (and some calculated) values for the electron affinities of the s- and p-block elements are presented in Table 2.3 from which a number of trends are apparent.

$$E(g) + e^- \rightarrow E^-(g) \tag{2.5}$$

We can note first of all that, with the exception of the noble gases, electron affinities generally increase (i.e. electron attachment becomes more exothermic) on crossing the p-block periods from left to right. This is the same trend as seen for ionization energies and for much the same reasons. Thus, as the effective nuclear charge increases, the energy released on adding an electron also increases. However, the trend is not regular most notably for the Group 15 elements which appear to have lower electron affinities than might be expected. The reason can be found by considering the electron configurations of the neutral atoms *vs* the anions formed and follows the explanation given in the previous section on ionization energies based around exchange energies.

In the case of the s-block elements, we should note two points. Firstly, the electron affinities for the Group 1 elements are larger than those for the corresponding Group 13 elements, and secondly, that the values for the Group 2

Table 2.3 Electron affinities (kJ mol^{-1}) for the s- and p-block elements

H 73							He <0
Li 60	Be −18	B 23	C 122	N −7	O 141	F 322	Ne <0
Na 53	Mg −21	Al 44	Si 134	P 72	S 200	Cl 349	Ar <0
K 44	Ca −186	Ga 36	Ge 116	As 77	Se 195	Br 324	Kr <0
Rb 47	Sr −146	In 34	Sn 121	Sb 101	Te 190	I 295	Xe <0
Cs 45	Ba −46	Tl 30	Pb 35	Bi 101	Po 186	At 270	Rn <0

Values in Table 2.3 are taken from Emsley (1989). All are experimental values except for gallium which is calculated. Values for the Group 18 elements are discussed below.

elements are all negative (i.e. endothermic). An explanation of the former observation lies in the fact that formation of a Group 1 anion results in a filled s orbital which has a certain stability; no such extra stability is gained by adding an electron to a Group 13 atom (although there is some gain in exchange energy). In the second case, addition of an electron to a Group 2 element places an electron in a higher-energy p orbital which is well shielded by the preceding ns^2 pair (see the explanation of why boron has a lower first ionization energy than beryllium).

In terms of group trends, we should note that the electron affinities for the 3p and 4p elements in the same group are similar as are those for the 5p and 6p elements (lead would seem to be an anomaly here). This reflects the larger effective nuclear charges found for the 4p and 6p elements as a result of the filling of the 3d and 4f rows respectively, a factor we referred to in Section 2.1 when considering a similar effect with ionization energies. This offsets the anticipated trend to lower values as the groups are descended.

Finally, we can note that the electron affinities for the first-row elements are generally lower than those for the second-row elements, particularly in the case of Groups 15, 16, and 17. This is a result of the very small sizes of the atoms of the first row elements, i.e. N, O, and F (see later), which is important, since the addition of an electron leads to a small negative ion with a high charge-to-size ratio resulting in large coulombic or interelectron repulsions. It is largely this size effect which leads to the first-row elements having lower than expected electron affinities.

The electron affinities for the noble gases are all calculated to be endothermic but are quoted as <0 in Table 2.3 since some uncertainty remains as to their actual values. Thus, although an outer electron in a Group 18 element feels a high effective nuclear charge, an extra electron must occupy an orbital of the next principal quantum shell which is further from the nucleus, and it will also be well shielded by the preceding s^2p^6 core.

2.4 Covalent and ionic radii

The concept of *atomic radii* whether they be covalent, ionic, or metallic must be considered carefully since we know from the discussion in Chapter 1 on atomic orbitals that an atom is not a hard sphere and that the electron density drops away exponentially as the distance from the nucleus increases. Nevertheless, such radii are very useful for explaining a lot of chemistry; we shall look briefly in this section at some of the observed trends in covalent and ionic radii.

A **covalent radius** is defined as half the length of a symmetrical, homonuclear element–element (E–E) bond (usually a single bond although we can calculate multiple-bond covalent radii as well), i.e. half of the E–E distance. It is particularly useful to have such radii available, since we can use them to calculate distances for previously unknown bonds. Thus, for example, if we calculate the covalent radius for A from A–A and that for B from B–B we should be able to estimate the bond length A–B. In general, this approach works well although we must be careful to take into account any electronegativity difference between the elements (see later) since any degree of ionicity or charge separation can lead to a shortening of the bond. A–B bond distances can be calculated from covalent radii more accurately by using the Shomaker–Stevenson relationship which explicitly corrects for any electronegativity differences between A and B.

Covalent radii for the s- and p-block elements are presented in Table 2.4, but it is important to keep in mind that the values quoted for covalent radii sometimes differ depending on the source consulted because their calculation is not always as straightforward and unambiguous as might be thought (we shall look at some reasons later). However, it is general trends in which we are interested, and those are clearly apparent from the values given in Table 2.4.

Two obvious trends may be discerned. Firstly, there is a decrease in covalent radius on moving from left to right across a period. Since all electrons are being added to the same principal quantum shell we might expect, other things being equal, that the atoms in any particular period would remain the same size. Of course, other things are not equal—most importantly, the effective nuclear charge which is increasing. The electrons are therefore more tightly held, the size of the orbital is contracted, and so the covalent radii are smaller.

Strictly speaking, a covalent radius is not a property of an isolated atom.

The Shomaker–Stevenson relationship takes the form $d_{AB} = r_A + r_B - C|\chi_A - \chi_B|$, where d_{AB} is the predicted bond length, r_A and r_B are the covalent radii in pm, C is a constant, and $|\chi_A - \chi_B|$ is the absolute difference in Pauling electronegativities of A and B. It should be noted that this is a purely empirical relationship and that its validity in certain circumstances has been questioned.

Fluorine is an example which illustrates that the determination of covalent radii is not always straightforward. Thus, a value for the covalent radius of fluorine of 0.54 Å rather than 0.64 Å has been suggested by Gillespie from a reappraisal of the bond lengths of a number of fluorine compounds.

Table 2.4 Covalent radii (Å) for the s- and p-block elements

H 0.30							He —
Li 1.23	Be 0.89	B 0.88	C 0.77	N 0.70	O 0.66	F 0.64	Ne —
Na 1.54	Mg 1.36	Al 1.25	Si 1.17	P 1.10	S 1.04	Cl 0.99	Ar —
K 2.03	Ca 1.74	Ga 1.25	Ge 1.22	As 1.21	Se 1.17	Br 1.14	Kr 1.10
Rb 2.47	Sr 1.92	In 1.50	Sn 1.40	Sb 1.41	Te 1.37	I 1.33	Xe 1.30
Cs 2.35	Ba 1.98	Tl 1.55	Pb 1.54	Bi 1.52	Po 1.53	At —	Rn —

Values of covalent radii in Table 2.4 are taken from Emsley (1989). Values for Na and Rb are quoted in this text as 'atomic radii'. Covalent radii are not available for some elements.

Secondly, there is a trend to increasing covalent radius as the groups are descended. This effect results from the fact that electrons are being placed in orbitals with increasing principal quantum number and therefore lie further from the nucleus. As with ionization energies and electron affinities, it is this factor which is of primary importance rather than any increase in effective nuclear charge. In fact, the observed values for atomic or covalent radii, r, of most elements are predicted quite well by Eqn. 2.6 (a_0 is the Bohr radius (53 pm), n is the principal quantum number and Z^* is effective nuclear charge) which clearly reveals the dominance of n on descending a group.

$$r = a_0 n^2 / Z^* \qquad (2.6)$$

Another point to note is the close similarity in size between the 3p and 4p elements. This is another consequence of the higher effective nuclear charge for the 4p elements resulting from the presence of the 3d row which serves to contract the orbitals and therefore make the elements smaller (note that gallium is the same size as aluminium). This is sometimes referred to as the 'scandide' contraction (from scandium) and is analogous to the better-known lanthanide contraction which results in the 5d elements being very similar in size to the 4d elements due to the presence of the 4f elements which precede the 5d row.

Ionic radii show much the same trends as covalent radii, and we will not dwell on the matter in much detail except to show that for the well-known halide anions, the radii increase regularly as the group is descended.

F^-, 1.33; Cl^-, 1.81; Br^-, 1.96; I^-, 2.20 (values in Å for 6-coordination)

Note that the ionic radii for the halide ions are significantly larger than the covalent radii for the corresponding halogen shown in Table 2.4. We should expect that radii will increase with increasing negative charge (or decreasing positive charge) due to increasing inter-electron repulsions, and a good illustration of this effect is seen by comparing the ionic and covalent radii for germanium and tin shown below. It is important to note that values chosen for ionic radii are very dependent on the **coordination number** (the value chosen here is 6), and we should not be surprised that covalent radii will also vary with coordination number often associated with a change in oxidation state and/or valence.

We should define coordination number at this point since we shall come back to it again in subsequent chapters. The coordination number is simply the number of nearest-neighbour atoms which surround the particular atom we are considering, and applies equally to covalent and ionic compounds.

	Radius (Å)		
	Covalent	+2 ionic	+4 ionic
Ge	1.22	0.87	0.67
Sn	1.40	1.12	0.83

A graph of covalent radii for the elements of the first two periods (excluding He and Ne) and hydrogen is shown in Fig. 2.2.

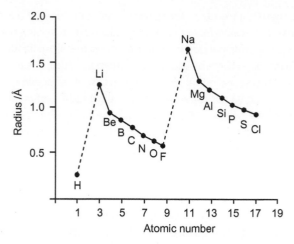

Fig. 2.2 Covalent radii for hydrogen and the elements of the first two short periods.

2.5 Electronegativity

We will see in the following discussion and in later chapters that ***electronegativity*** can be used as an extremely powerful organizing principle for understanding the nature of the elements themselves and the types of compound they form with each other. It has even been argued (by Allen, see below) that electronegativity should be considered as the third dimension of the periodic table.

The concept of electronegativity (hereafter referred to with the symbol χ) had certainly been recognized by Berzelius more than 150 years ago, but it was Pauling in the 1930s who provided a quantitative scale together with a simple and concise definition: ***the ability of an atom to attract electron density towards itself in a molecule***. The Pauling scale of electronegativity, traditionally quoted on a scale of 0–4, is still one of the most widely used by chemists, and a compilation of Pauling electronegativity values (χ_P) for the s- and p-block elements is given in Table 2.5. Other commonly used scales are those of Allred and Rochow, in which the element electronegativity is defined as proportional to Z^*/r^2 (where r is the covalent radius), and the Mulliken scale where χ is proportional to half the sum of the first ionization energy and the first electron affinity (χ_M; we shall briefly meet this one again in Chapter 6).

This is not the place to discuss the extensive literature on the topic of electronegativity in great detail (and it is very extensive), particularly the merits of the many different scales which have been developed over the years, but it will be instructive to comment on the work of Allen in this area (*J. Am. Chem. Soc.*, 1989, **111**, 9003). It is Allen's contention that the electronegativity of an element is ***the average one-electron energy of valence shell electrons in the ground state***

Elements described as ***electronegative*** have high values of electronegativity (e.g. fluorine), and elements described as ***electropositive*** have low values of electronegativity (e.g. sodium).

free atoms and that it can be determined from experimental spectroscopic data according to the Eqn. 2.7, where ε_p and ε_s are the p and s ionization energies and *m* and *n* are the number of p and s electrons. Electronegativity values calculated according to this formula, χ_{spec}, and adjusted to the Pauling scale, are presented in Table 2.6.

$$\chi_{spec} = \left(m\varepsilon_p + n\varepsilon_s\right)/(m + n) \tag{2.7}$$

Table 2.5 Pauling electronegativities for the s- and p-block elements

H 2.20								He —
Li 0.98	Be 1.57		B 2.04	C 2.55	N 3.04	O 3.44	F 3.98	Ne —
Na 0.93	Mg 1.31		Al 1.61	Si 1.90	P 2.19	S 2.58	Cl 3.16	Ar —
K 0.82	Ca 1.00		Ga 1.81	Ge 2.01	As 2.18	Se 2.55	Br 2.96	Kr —
Rb 0.82	Sr 0.95		In 1.78	Sn 1.96	Sb 2.05	Te 2.10	I 2.66	Xe —
Cs 0.79	Ba 0.89		Tl 2.04	Pb 2.33	Bi 2.02	Po (2.0)	At (2.2)	Rn —

Values in Table 2.5 are taken from Emsley (1989). The reason for the 0–4 scale stems from the fact that the original calculations derived from bond energy data which were quoted in eV. 1 eV = 96 kJ mol⁻¹.

Table 2.6 Allen electronegativities calculated for the s- and p-block elements

H 2.300								He 4.157
Li 0.912	Be 1.576		B 2.051	C 2.544	N 3.066	O 3.610	F 4.193	Ne 4.787
Na 0.869	Mg 1.293		Al 1.613	Si 1.916	P 2.253	S 2.589	Cl 2.869	Ar 3.242
K 0.734	Ca 1.034		Ga 1.756	Ge 1.994	As 2.211	Se 2.424	Br 2.685	Kr 2.966
Rb 0.706	Sr 0.963		In 1.656	Sn 1.824	Sb 1.984	Te 2.158	I 2.359	Xe 2.582
Cs 0.659	Ba 0.881		Tl 1.789	Pb 1.854	Bi 2.01	Po 2.19	At 2.39	Rn 2.60
Fr 0.67	Ra 0.89							

Values in Table 2.6 (with the exception of those for Fr and Ra) are taken from the IUPAC Technical Report on Oxidation States authored by Karen, McArdle and Takats (*Pure Appl. Chem.*, 2014, **86**, 1017).

A slightly simpler approach to calculating element electronegativities, but one based very much on the method proposed by Allen, has been described by Gillespie and co-workers (*J. Chem. Ed.*, 1996, **73**, 627).

Note that the Allen electronegativities for Fr and Ra are slightly higher than the values for Cs and Ba respectively consistent with the slightly higher first ionization energies of these 7s elements noted in Section 2.2.

For the most part, the Pauling values (χ_P) and χ_{spec} are rather similar, and we will not look at any comparisons in detail except to note that whereas the Pauling scale indicates that chlorine is more electronegative than nitrogen, the Allen scale suggests that the reverse is true. The relative electronegativities of these two elements has been a longstanding matter of uncertainty in inorganic chemistry. Allen has also made the point, referred to earlier, that we can think of electronegativity as being the third dimension of the periodic table, and this is shown graphically in Fig. 2.3.

This three-dimensional view allows us to see immediately the trends in electronegativity for the s- and p-block elements as a whole. Thus, if we look at the trends by row, as shown specifically in the graph in Fig. 2.4, it is apparent that electronegativity increases fairly regularly across the periods with the rate of increase tending to decrease as the s- and p-blocks are descended. This regular trend of increasing electronegativity across a period mirrors the trend in effective nuclear charge, but we see none of the kinks that we saw in the values for ionization energies or electron affinities (note that electron affinity and electronegativity are not the same thing). This should not be surprising, however.

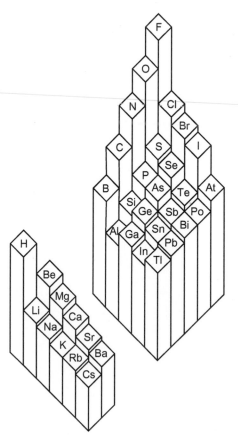

Fig. 2.3 A view of the periodic table for the s- and p-block elements (excluding those in Group 18) showing electronegativity (using Pauling values) as a third dimension.

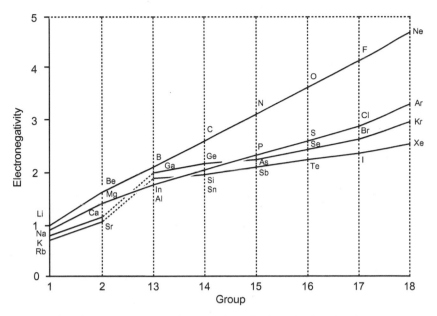

Fig. 2.4 Allen electronegativities as a function of row for the 2s/2p to 5s/5p elements. Dotted lines between Ca and Ga and between Sr and In indicate the presence of 3d and 4d elements respectively.

Ionization energies and electron affinities are properties of isolated, individual atoms whereas electronegativity (even though on some scales, measurable from the properties of isolated atoms) is a property of atoms in molecules. The stabilities of certain electronic configurations important in isolated atoms are no longer relevant for atoms in molecules.

In Fig. 2.5, a graph is shown of Allen electronegativities according to group from which a number of other trends are apparent. Firstly, there is a general decrease in electronegativity as the groups are descended, but the similarity in the values for the 3p and 4p elements is clear. Moreover, in the case of Groups 13 and 14, the electronegativity of the 4p element is actually greater than that of the 3p element, i.e. Ga > Al and Ge > Si. This is a further example of the consequences of the increased effective nuclear charges of the 4p elements resulting from the presence of the 3d row.

We shall have more to say on the electronegativities of the elements in Chapter 3 when we look at the properties of the elements themselves and where the meaning of the metalloid band in Fig. 2.5 will be addressed. Remember, however, that the values of χ_{spec} discussed previously are for neutral, isolated atoms. In compounds of the elements, the precise values will depend on a number of factors such as oxidation state and hybridization as well as the other atoms to which the element in question is bonded. We will look at some of these factors in more detail later, but in terms of oxidation state we may note here, for example, that the electronegativity of thallium according to the Pauling scale is considerably larger in thallium(III) compounds ($\chi_P = 2.04$) than in those of thallium(I) ($\chi_P = 1.62$), the same being true for lead (2.33 for Pb(IV) vs 1.87 for Pb(II)); values

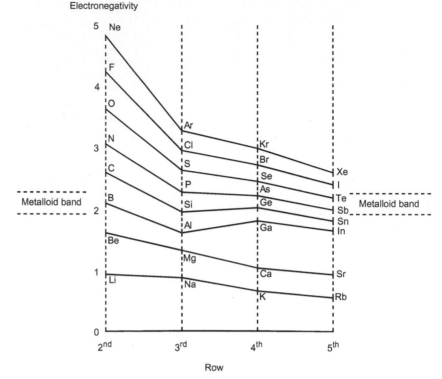

Fig. 2.5 Allen electronegativities for the 2s/2p to 5s/5p elements as a function of group.

Whilst nitrogen is more electronegative than carbon, carbon has the higher electron affinity. As noted in the main text, an explanation lies in the fact that electron affinities are thermodynamic quantities for free atoms for which exchange energies and so on are important. Because of the way they are calculated, electronegativity values (whether Allen, Pauling, or others) do not reflect the stability of certain electron configurations and therefore show a smooth increase across a period in line with increasing effective nuclear charge.

A discussion of hybrid orbitals is given in Winter (2016).

quoted in Table 2.5 are for the higher element oxidation state where differences are significant.

With regard to hybridization, it has long been recognized that, in the case of carbon, the order of electronegativity is sp > sp² > sp³, the reason being that electrons are more tightly held (lower in energy) in s orbitals, and more advanced treatments of electronegativity assign specific electronegativities to particular orbitals or 'valence states' of atoms. In terms of the other atoms to which the element in question is bonded, the phosphorus atom in a PF_2 group, for example, is more electronegative than the phosphorus in a PMe_2 group reflecting the greater electronegativity of F *vs* Me. This introduces the concept of group electronegativity, but we shall not dwell on the details of that here. Finally, electronegativity is sometimes regarded as a chemical potential of an atom which is often a useful guide to reactivity; we will return to this matter in Chapter 6.

2.6 Orbital energies and promotion energies

Orbital energies and the energies of electron promotion from the ground state to a so-called 'valence state' are important atomic properties and are useful in understanding certain aspects of periodicity, for example, in discussions about the so-called inert pair effect which we will look at in more detail later. We will consider some of the trends in this section.

Fig. 2.6 shows the calculated one-electron energies of the valence s and p orbitals for the p-block elements. We can note first that within a particular group, the orbital energies increase (become less negative) as the group is descended (i.e. electrons become easier to remove), and also that, in general, the s-p energy separation decreases. The only exception to this generalization is with some of the elements of the 4p row. Thus, Ga and Ge, in particular, have larger s-p separations than Al and Si respectively, with the energy of the s orbitals lower than their congeners in the 3p row. Moreover, for the remaining elements, As, Se, Br, and Kr, the s orbital energy is lower than would be anticipated on the basis of the gradual trends expected from elements in other rows. This observation is another consequence of the increased effective nuclear charge of the 4p elements following the 3d row discussed previously, and it is clear that the effect on s orbitals is larger than that on p orbitals which is due to the greater penetration of s orbital electron density into the inner electron core. The parallel with trends in electronegativity is obvious; indeed, the similarity with the Allen scale is a result of the fact that this scale is related to orbital energies.

We might also expect a similar situation for the 6p elements as a result of the presence of the 4f row and relativistic effects. If the effect is present at all, it is clearly not as dramatic as for the 4p row, but the s-p separations for Tl and Pb are very similar to those of In and Sn respectively; for the later 6p elements it is not so obvious.

One-electron orbital energies are often used with an underlying assumption that orbital energies do not change according to orbital occupation (Koopmans' theorem) but this is an oversimplification. Thus, for polyelectronic atoms, orbital energies are *not* equal to ionization energies, since ionization energies reflect not just the energy required to remove an electron but also the effect of this removal on the energies of the remaining electrons.

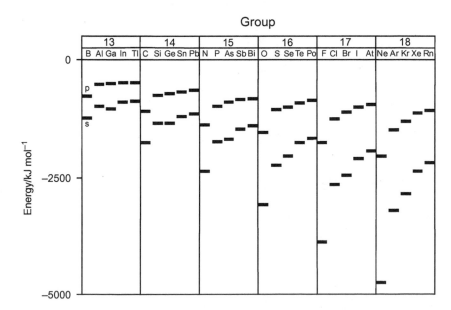

Fig. 2.6 Calculated one-electron energies of the valence s and p orbitals for the p-block elements. For each element, the s orbital is lower in energy than the p orbitals as shown explicitly for boron.

The other obvious trend is that both s and p energies decrease markedly as a period is crossed from left to right, particularly for the first-row (2p) elements, and also that the s-p separation increases considerably. This reflects the large increase in effective nuclear charge on moving from left to right, and the more pronounced effect on s orbitals is also apparent.

One of the problems with one-electron orbital energies is that they do not take account of so-called electron correlation effects, i.e. the effect that moving an electron from one orbital to another has on the energies of other electrons (which is why ionization energies and orbitals energies are not the same as noted in the margin on the previous page). This reorganization energy, as it is sometimes called, can be quite substantial, particularly for the heavier elements, and is revealed in the experimentally obtained electron promotion energies. Fig. 2.7 shows the promotion energies for the Group 13 and 14 elements, these being the energies required for the electronic transitions ns^2np^1 to ns^1np^2 and for ns^2np^2 to ns^1np^3 respectively. Perhaps the most notable and somewhat unexpected feature is that the promotion energies of the 6p elements Tl and Pb are the highest in their groups indicating that the reorganization energy is considerable. We will return to a discussion of this matter in Chapter 4.

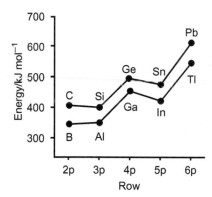

Fig. 2.7 Experimental electron promotion energies for the elements of Group 13 and 14. The actual values (in kJ mol^{-1}) are B, 346; Al, 349; Ga, 456; In, 420; Tl, 543; C, 405; Si, 400; Ge, 504; Sn, 476; Pb, 609; taken from Dasent (1965).

2.7 Relativistic effects

One of the consequences of the theory of special relativity is that objects moving close to the speed of light increase in mass. Thus, when the velocity (v) of an electron in an atom approaches the speed of light (c), its mass (m) increases according to Eqn. 2.8 where m_0 is the rest mass of the electron.

$$m = m_0\left[1-\left(v/c\right)^2\right]^{-1/2} \tag{2.8}$$

In order to appreciate the importance of this relativistic increase in electron mass, we should note two further points. In Eqn. 2.1 ($E_n = -(Z^{*2}R_H)/n^2$)

the Rydberg constant R_H is actually a collection of other constants which can be expressed as shown in Eqn. 2.9.

$$R_H = m_e e^4 / 8\varepsilon_0^2 h^2 \qquad (2.9)$$

The energy of the electron in an atom is therefore dependent on its mass (m_e in Eqn. 2.9; we will not define all the other terms) such that it will decrease in energy (become more stable) as its mass increases. Also, it is not just the energy of the electron which is affected. The maximum in the radial probability function (the Bohr radius a_0) is also affected according to Eqn. 2.10 such that the radius becomes smaller with increasing electron mass (a_0 is inversely proportional to m_e; again, we will not define the other terms here).

$$a_0 = \varepsilon_0 h^2 / m_e \pi e^2 Z \qquad (2.10)$$

Why all this matters is that for electrons in the heavier elements of the periodic table, electron velocities can be an appreciable fraction of the speed of light as a result of the high nuclear charge, and these electrons are therefore subject to so-called **relativistic effects**. (Strictly speaking, electrons in all atoms are subject to relativistic effects, but the magnitude of the effect and its consequences are only appreciable for the heavier elements.) Thus, as the velocity, and hence mass, of the electron increases, the effective size of the orbital decreases (**direct relativistic orbital contraction**) and its energy is lowered. The effect is most pronounced for electrons in s orbitals, since it is these orbitals which have the greatest density near the nucleus (recall that in graphs of ψ and ψ² vs r, described in Chapter 1, only s orbitals have non-zero values at the nucleus), rather less marked for p orbitals and relatively unimportant for d and f orbitals, although these latter two types of orbital are affected indirectly by the contraction of the s and p orbitals such that they are better shielded and hence more diffuse than they would otherwise be (**indirect relativistic orbital expansion**). For the heavier p-block elements Tl through Rn (the 6p elements) there is an appreciable relativistic stabilization of the electrons in the 6s orbital which has important consequences for the chemistry of these elements as we shall see in Chapter 4. In fact, when relativistic effects are taken into account, the energy of the 6s orbital for the 6p elements (Tl to Rn) is found to be lower than the 5s orbital for the 5p elements (In to Xe), a feature not manifest in Fig. 2.6 which is based on non-relativistic energy calculations.

We shall conclude this section by noting that it was Dirac who was the first to consider the importance of relativistic effects in terms of electrons in atoms. The Dirac equation is essentially a relativistic treatment of the Schrödinger equation, the details of which will not detain us here except for one important result. For the heavier or high Z elements, i.e, the 6p elements as far as this text is concerned, the orbital angular momentum and spin angular momentum, described by the quantum numbers l and s (see Chapter 1), can no longer be treated as separate. So-called spin–orbit coupling becomes increasingly important such that a new quantum number $j = l + s$ is required. One of the consequences of this coupling is that the three p orbitals in a given shell are no longer degenerate but split into one $p_{1/2}$ and two $p_{3/2}$ orbitals (for reasons we will not go into here), the former

It is important to stress that although relativistic effects are often described as being a significant feature of the 'heavier' elements, it is not the mass of the element which is important but its high nuclear charge. We might better say, therefore, that relativistic effects becoming increasingly important for high atomic number or high Z elements.

In the case of the 1s electron in bismuth, for example, the electron reaches a speed of around 60% c which results in a mass of 1.26 m_0 resulting in a reduction in the radial extension (Bohr radius) of the 1s orbital by about 20%.

Spin–orbit coupling effects are important in explaining the trends seen in Fig. 2.8, particularly in terms of the large promotion energies seen for Tl and Pb.

It may seem strange to be talking about electron velocities and, perhaps by implication, of electrons as particles after all the discussion in Chapter 1 on electrons as waves. However, although it may be easier to think about relativistic effects in terms of fast-moving particles, all of the consequences of relativistic effects emerge from a wave model in just the same way, albeit in a manner which may be less intuitively obvious.

one being of lower energy. As we will briefly note in Chapter 4, this splitting of the p orbitals is predicted to be important for the chemistry of the newly identi-fied 7p elements.

Exercises

1. Using Slater's rules, calculate the values of S and Z^* for an outermost electron in the elements B, F, Mg, and P.

2. Considering the electronegativity of sp, sp^2, and sp^3 hybrid orbitals noted in Section 2.5, account for why terminal alkynes are more readily deprotonated than alkanes.

3. What is the difference between electron affinity and electronegativity?

3 Periodicity in the properties and structures of the elements

3.1 General points

In this chapter we will look at the properties of the elements themselves and attempt to understand the origin of the observed periodic trends. The first, and perhaps most important, point to note is that the p-block is the only part of the periodic table which contains non-metals; all of the s-, d-, and f-block elements are metals (although see the marginal note with regard to hydrogen). Moreover, a look at Fig. 3.1 reveals that within the p-block, the metals are found in the bottom left-hand corner whilst the so-called metalloids or semi-metals, shown in circles, lie along a diagonal from top left to bottom right with the non-metals, shown in bold type, occupying the top right. There is, therefore, a trend from metallic to non-metallic character as we move from bottom left to top right. This pattern correlates precisely with the trend seen for the element electronegativities discussed in Chapter 2 and, as we shall see, it is electronegativity which we can use to rationalize the observations described above.

Before we look for an explanation of why particular elements have the properties they do, it is as well to be clear about what is meant by the terms 'metal, metalloid, and non-metal'. For our purposes, it is probably sufficient to concentrate on the electrical properties whereby we can define metals as conductors, non-metals as insulators, and metalloids as semiconductors. These properties can be understood in terms of the so-called **band theory** of solids which we can illustrate with the aid of the simple diagrams shown in Fig. 3.2.

Hydrogen is something of a special case. It is certainly a non-metal under normal conditions and its electronegativity (2.20, Pauling; 2.300, Allen) is consistent with its being placed in the non-metal class. It is therefore somewhat different from the other s-block elements which is why it is sometimes placed in group on its own (often with helium) as noted in Section 1.6 in Chapter 1.

H							He
Li	Be	(B)	**C**	**N**	**O**	**F**	**Ne**
Na	Mg	Al	(Si)	**P**	**S**	**Cl**	**Ar**
K	Ca	Ga	(Ge)	(As)	**Se**	**Br**	**Kr**
Rb	Sr	In	Sn	(Sb)	(Te)	**I**	**Xe**
Cs	Ba	Tl	Pb	Bi	Po	**At**	**Rn**

Fig. 3.1 The s- and p-block part of the periodic table illustrating the regions of metals, metalloids, and non-metals; non-metals in bold, metalloids in circles

The categories illustrated in Fig. 3.1 are not without some ambiguity particularly with regard to which elements are classed as metalloids; borders are often open to interpretation. Thus, bismuth and polonium are sometimes described as metalloids.

Metallic conductivity is not a feature only of metals (or alloys); it is also observed in some solid-state compounds, as we shall see in later chapters, as well as in graphite (see below).

Band theory is essentially the molecular orbital theory of solids. For small molecules, we get a small number of discrete molecular orbitals equal to the number of atomic orbitals from which they are derived. For solids the number of atomic orbitals is very large, resulting in groups of molecular orbitals which are very closely spaced in energy called bands. More on band theory can be found in Hoffmann (1988).

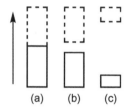

Fig. 3.2 Diagrams showing the band structure of (a) a conductor, (b) a semiconductor, and (c) an insulator. Orbital or band energy increases in the direction of the arrow.

For the s- and p-block metals in particular, the number of valence electrons is generally less than the number of valence orbitals which necessarily results in the bands being only partially filled.

For conductors (Fig. 3.2(a)) we have a partially filled band of orbitals (the bold lines indicate that part which is filled, the dashed lines the part which is unfilled), and electrons can readily be excited into unfilled levels in which they can move easily through the solid. An important characteristic is that the resulting high electrical conductivity decreases slightly with increasing temperature (due to increased scattering of electrons resulting from lattice vibrations). For semiconductors (Fig. 3.2(b)), a filled band is separated from an unfilled band by a small energy gap, the so-called band-gap. Electrons are easily promoted by thermal excitation (according to the Boltzmann distribution) from the filled levels (in which they cannot move) to the unfilled, delocalized, so-called conduction band resulting in a dramatic increase in conductivity with temperature (this outweighs the tendency for conductivity to decrease with temperature as found in metals). For insulators (Fig. 3.2(c)), the band-gap is too large to allow for easy promotion of electrons and so no conduction occurs. It is, of course, based on the definitions above, somewhat arbitrary where semiconductors end and insulators begin.

Returning now to the elements, a comparison of Fig. 3.1 and the Allen electronegativities shown in Table 2.6 in Chapter 2 reveals that elements with an electronegativity smaller than about 1.9 are metals whereas those with electronegativities greater than about 2.2 are non-metals. Thus, the metalloids all fall within the fairly narrow range of electronegativity of 1.9 to 2.2, and we can use this range of electronegativities to define the metalloid band which is shown explicitly in Fig. 2.5. Let us see why it is that electronegativity is important in this regard.

In Chapter 2 we saw that the energy separation between the valence s and p orbitals is related to electronegativity such that those elements with the largest electronegativity also have the largest valence s-p energy separation. Thus, if we consider the elements with low electronegativity, the s and p orbitals are close together in energy and are therefore able to easily mix or overlap with each other (this is true whether we are considering s-p mixing on a particular atom, i.e. hybridization, or s-p overlap between atoms in the formation of molecular orbitals). The result of this ease of mixing is that when a number of atoms are placed in close proximity to each other, there is extensive orbital overlap between atoms (s-s, p-p, and s-p) and closely spaced energy bands with a relatively large width are formed. This overlap is made easier by the fact that orbitals of the elements of low electronegativity tend to be fairly diffuse with correspondingly large radial extensions since the electrons are not tightly held. Wide and closely spaced bands tend to overlap with each other resulting in one large band which is only partly filled with electrons. This is a characteristic of metals wherein the bonding is highly delocalized, and the fact that they are only partly filled results, as we have seen previously, in electrical conductivity.

For elements with a large electronegativity, the s-p energy separation is correspondingly large. Extensive s-p mixing and overlap between the orbitals of neighbouring atoms is therefore reduced with the result that extensively delocalized bonding is disfavoured with respect to the formation of localized covalent bonds. For the elements nitrogen, oxygen, and fluorine this includes multiple bonding and covalently bonded diatomics result whereas for elements such as phosphorus and sulfur, single bonds predominate which result in larger molecular or polymeric structures. In these elements there are large energy separations between filled and unfilled orbitals or bands even for polymeric structures, and as a consequence they are insulators. The fact that high electronegativity results in valence electrons being tightly bound and therefore in relatively smaller orbitals is also a reason for why the delocalized bonding found in metals is unfavourable for these high-electronegativity elements.

The situation for the metalloids is intermediate between these two extremes. The band structure of the polymeric or macromolecular structures which these materials adopt resembles that of the non-metals in that bands are either filled or unfilled, but the band gap is small and electrons are therefore easily promoted to the conduction band resulting in electrical conductivity.

An alternative rationale for the observed metallic or non-metallic properties of the elements is associated with what is called a Peierls distortion. Briefly, a Peierls distortion (the solid-state analogue of a first-order Jahn–Teller distortion seen in molecules where degenerate or nearly degenerate orbitals are only partially filled with electrons) is expected to occur in any situation where there is a partially filled band resulting in the formation of a band-gap and an overall lowering of the energy of the filled orbitals at the expense of raising the energy of the unfilled ones. From a structural point of view, this results in a delocalized structure distorting to a structure of lower symmetry in which the bonding is more localized.

A more detailed discussion of Peierls distortions and Jahn–Teller distortions (first and second order) is given in Albright, Burdett, and Whangbo (1985) and also in Hoffmann (1988).

In principle, any structure with a partially filled band, i.e. a metal, should distort, but the magnitude of the distortion is all-important. This magnitude depends on the number of electrons present (the more you can stabilize, the better) but also, crucially, on the element electronegativity and its relation to valence orbital energies. Thus, if we start with a hypothetical delocalized structure, we can see what happens as we move from left to right in the periodic table. On the left-hand side where electronegativity is small, all the orbitals are close in energy and there is not much to be gained from lowering the energy of some of them, and this factor, coupled with the small number of electrons present, means that there is little driving force for any distortion to a structure with more localized bonding, and a metallic structure is preferred. Moreover, any small distortion that might be favoured energetically is likely to be overcome by thermal vibrations in the lattice and/or interatomic repulsive terms. For elements on the right, however, the fact that the valence orbital energies are widely separated means that a lot can be gained by distorting a hypothetical delocalized structure to a more localized one since it is possible to lower some orbitals in energy by a considerable amount. This factor, coupled with the larger number of valence electrons present, results in a large driving force for distortion to a structure with localized rather than delocalized covalent bonding, i.e. a non-metallic form.

A final point which sheds a little more light on metallic structures and why they are favoured can be made by considering lithium as an example. In the gas phase, lithium exists as Li_2 molecules with a bond dissociation energy of about 100 kJ mol^{-1}. A molecular orbital energy level diagram (see, Winter (2016)) reveals that for Li_2, low-energy unfilled bonding orbitals derived from the overlap of the 2p orbitals are present, and in situations like this (i.e. low-energy unfilled bonding orbitals) there is a tendency for some other arrangement of atoms to be favoured which essentially works to fill these orbitals with electrons. The new arrangement in this case (and many others) is a metallic structure (which we shall look at in detail later) with extensive delocalized bonding, and the energetic advantage of this is clear when we compare the dissociation energy of Li_2 noted previously with the atomization energy of lithium metal which is about 140 kJ mol^{-1}, i.e. substantially greater.

It will now be useful to look at the structures of the elements in more detail in order to better appreciate some of the points we have just discussed although we shall not deal with this topic by way of a comprehensive discussion of all the known **allotropes** which, for some elements, is quite extensive.

The term **allotrope** is used here to distinguish between different structural forms of the same element; for example, white phosphorus, P_4, and polymeric red phosphorus. **Polymorphism (polymorphs)** is a term used to distinguish between different crystalline modifications of the same molecular form such as the orthorhombic and monoclinic forms of S_8.

3.2 Groups 1 and 2

With the exception of hydrogen, which exists under normal conditions as a gas containing diatomic molecules, all the elements of Groups 1 and 2 are metals and have typical metallic structures. All of the Group 1 elements (Li to Cs) adopt a body-centred cubic structure (which is not quite close-packed) as do barium and radium in Group 2. Beryllium and magnesium have hexagonal close-packed structures whereas calcium and strontium are face-centred cubic (i.e. cubic close-packed). In many cases, elements adopt a different structure when the temperature and/or pressure is changed.

There are two types of close-packed structures: namely hexagonal close-packed and cubic close-packed (also known as, and usually called, face-centred cubic). These differ in the packing arrangements of atoms with layers ABABAB for hexagonal close-packing and ABCABC for cubic close-packing; in each case every atom has twelve nearest neighbours. The body-centred cubic structure is a more open structure which is not quite close-packed where each atom has eight nearest neighbours. For more detail, see Keeler and Wothers (2008).

3.3 Group 13

The only non-metallic element in this group is boron, which is generally classed as a metalloid. Elemental boron exists as a number of different allotropes, all of which are polymeric or macromolecular solids (some **crystalline**, some **amorphous**), the most well-defined being α- and β-rhombohedral boron. Both of these allotropes contain B_{12} icosahedra, linked together in three-dimensional arrangements. α-rhombohedral boron (Fig. 3.3(a)) can be viewed as a cubic close-packed array of B_{12} icosahedra whereas β-rhombohedral boron has a more complex structure with a central B_{12} unit at the heart of a B_{84} cluster which forms the repeat unit of this allotrope. This structural complexity is a manifestation of the fact that boron is **electron deficient** in the sense that it has four valence orbitals but only three valence electrons. The bonding between boron atoms is therefore mostly of the multicentre type rather than the more usual two-centre, two-electron bonds found for **electron precise** structures for elements further to the right in the periodic table. We should note that the electronegativity of boron places it firmly in the metalloid band and that the elemental structures, complex though they are, nevertheless exhibit a degree of bond localization not found in typical, fully delocalized metallic structures. We shall see something slightly similar for gallium below.

The use of the Greek letter labels α- and β- to distinguish between different allotropes or polymorphs is commonplace.

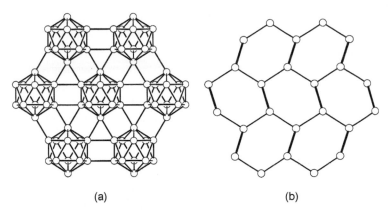

(a)   (b)

Fig. 3.3 (a) Part of the structure of α-rhombohedral boron and (b) part of the structure of elemental gallium with the short Ga–Ga distances highlighted.

The remaining elements in this group are all metals. Aluminium and thallium each have typical close-packed metallic structures with twelve nearest neighbours (cubic close-packed and hexagonal close-packed respectively). Indium has a slightly distorted cubic close-packed structure (body-centred tetragonal) with four atoms marginally closer than the other eight. Gallium, however, has a rather unusual structure (Fig. 3.3(b)) in which each gallium atom has one neighbour (2.44 Å) considerably closer than the others which occur in pairs at 2.70, 2.73, and 2.79 Å. In fact, the structure of solid gallium has some similarities to that of solid iodine, which consists of discrete I_2 molecules (see later), and it has been suggested that weakly bonded Ga_2 molecules are present in the solid phase of this element.

Group 13 is often considered to be the worst group in the periodic table for trends since, inasmuch as any are apparent, they often seem to be rather irregular. Note, however, that we can use electronegativity to rationalize some of what is outlined above. Boron, as we have said, is too electronegative to be a metal, but the least electronegative elements—aluminium, indium, and thallium—adopt typical metallic structures albeit slightly distorted in the case of indium. Gallium, the second most electronegative element in the group, has a structure in which incipient bond localization (in the form of the Ga_2 units mentioned earlier) typical of the metalloids is present.

Terms such as cubic, hexagonal, rhombohedral, tetragonal, orthorhombic, and monoclinic mentioned previously and below refer to the symmetry of the crystal structure. Definitions and diagrams can be found in any of the texts on solid-state chemistry listed in the bibliography.

3.4 Group 14

Group 14 is perhaps the group that best illustrates the trend from non-metallic to metallic properties as it is descended reflecting the corresponding decrease in element electronegativity. Carbon is a non-metal and is found in two common allotropic forms, namely diamond and graphite, the structures of which are illustrated in Figs. 3.4(a) and 3.4(b). In diamond, each carbon atom is surrounded by four nearest neighbours at the vertices of a tetrahedron, the atoms being linked by strong, directional, two-centre, two-electron covalent bonds.

Graphite, in contrast, is a layered structure consisting of planar hexagonal sheets stacked on top of each other and, under ambient conditions, is the thermodynamically most stable carbon allotrope. Each carbon atom in graphite is surrounded by three nearest neighbours, and extensive element–element π-bonding is important, resulting from the overlap of one p orbital per carbon atom aligned perpendicular to the hexagonal sheet. It is this feature that gives rise to graphite being a conductor and, in this regard, makes it something of an exception as a non-metal.

Both diamond and graphite occur as two polymorphs. Cubic diamond is the most common (Fig. 3.4(a)) in which all six-membered rings have a chair conformation whereas in the other form, hexagonal diamond, some rings adopt the slightly less stable boat form. The most stable form of graphite is the hexagonal form (sometimes called α-graphite) shown in Fig. 3.4(b), in which the carbon atoms in every other layer lie vertically above each other, i.e. the stacking is ABABAB. A rhombohedral modification (β-graphite) is also known in which the planar sheets adopt an ABCABC stacking arrangement. This second form has an interesting relationship to the cubic diamond structure, a feature to which we shall return later.

> More detail on chair and boat forms of six-membered rings can be found in any standard text on organic chemistry.

Extensive π-bonding is also a feature of the structures of the molecular forms of carbon, the fullerenes, of which the archetype is Buckminsterfullerene, or C_{60}, shown in Fig. 3.4(c). Many other fullerenes of higher molecular weight are known, e.g. C_{72}, C_{76}, C_{84} together with carbon nanotubes (the structural variety of which is considerable), as well as graphene, which comprises single layers of graphite.

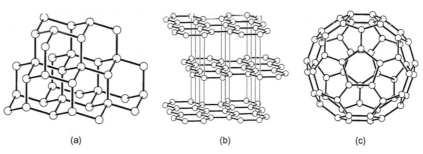

(a)   (b)   (c)

Fig. 3.4 (a) Part of the structure of cubic diamond, (b) part of the structure of hexagonal graphite, and (c) the structure of Buckminsterfullerene, C_{60}.

> Another difference between diamond and silicon or germanium is that pure diamond is clear and colourless (colours in natural or synthetic diamonds are due to impurities) whereas silicon and germanium are black. These properties also result from the differences in the band-gaps; visible light has sufficient energy to promote electrons from the lower to the higher band in silicon or germanium (but not in diamond) and hence is strongly absorbed.

Silicon and germanium both have the cubic diamond structure and are classified as metalloids largely as a result of their semiconducting properties. These are elements of intermediate electronegativity in which localized covalent bonding occurs but for which the valence s-p energy separations are small enough to result in fairly closely spaced bands. Diamond, which has a similar band structure to Si and Ge, is generally classed as an insulator but only because of its much larger band gap resulting from the greater s-p energy separation. The larger band gap in diamond is also a result of the smaller size of carbon; the C–C bonds are therefore significantly shorter than the corresponding Si–Si and Ge–Ge bonds which leads to correspondingly better orbital overlap.

Tin lies close to the metalloid-metal border and exists as two allotropes, grey and white tin—the former, α-Sn, having the cubic diamond structure.

White tin (β-Sn) has a less regular structure, still with four nearest neighbours (slightly further away than in grey tin), but with additional tin atoms much closer than the second nearest neighbours in the grey tin structure making white tin much denser and more metallic in its properties. Lead adopts a close-packed structure (cubic close-packed in this case) consistent with its metallic character as expected from its low electronegativity.

3.5 Group 15

Nitrogen is a true non-metal and exists exclusively as a diatomic molecule, N_2, in which the two nitrogen atoms are held together by an extremely strong triple bond (Fig. 3.5(a)). This directional, localized covalent bonding is to be expected for such an electronegative element, and we should also note the presence and importance of N–N π-bonding.

In contrast to nitrogen, phosphorus does not exist as P_2 molecules except under conditions of high temperature and low pressure in the gas phase. The stable elemental forms are white (or yellow) phosphorus, which contains tetrahedral P_4 molecules (Fig. 3.5(b)), and a variety of polymeric allotropes such as red, black, and violet phosphorus. All forms of phosphorus contain trigonal pyramidal, three-coordinate phosphorus atoms in which each phosphorus is linked to three others by single P–P bonds. This feature illustrates an important difference between the chemistry of phosphorus and nitrogen, and between the first- and second-row elements generally, in that there is a reluctance of the heavier elements to form stable element–element π-bonds resulting in a prevalence of element–element single bonding or **catenation**—a topic which we will discuss in much more detail in Chapter 4. The structure of orthorhombic black phosphorus, the most thermodynamically stable elemental form, is illustrated in Fig. 3.5(c) and consists of layers of linked six-membered rings in the chair conformation, whilst that of a closely related higher-pressure rhombohedral form is shown in Fig. 3.5(d). At even higher pressures, a simple cubic structure analogous to

The high-pressure cubic form of phosphorus is an exception to the general point made that all phosphorus atoms are trigonal pyramidal in phosphorus allotropes; we shall return to this structure in Section 3.9. It is also metallic, which, as noted previously, is unusual for a non-metal but this is closely associated with its structure.

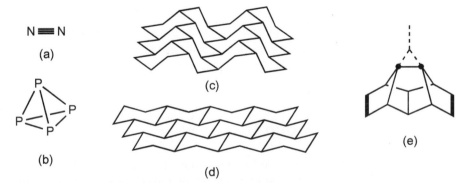

Fig. 3.5 (a) Dinitrogen, N_2, (b) tetraphosphorus, P_4, (c) part of a layer in orthorhombic black phosphorus, (d) part of a layer in rhombohedral black phosphorus, and (e) a unit found in red and violet phosphorus.

α-polonium (see later) is adopted in which each phosphorus is surrounded by six others in an octahedral coordination geometry.

Red phosphorus is known in both amorphous and crystalline forms, and violet phosphorus is another known crystalline allotrope. Fig. 3.5(e) shows a unit that is apparently common to all red and violet forms. Ribbons are formed by linking these units where the bold bonds are indicated (i.e. these indicated bold bonds are shared between adjacent units) whereby each unit can be in the same orientation as its neighbour or the other way up such that orientations alternate along the ribbon. The two phosphorus atoms represented as filled black circles in Fig. 3.5(e) can be linked as shown with the solid bond or in an arrangement indicated by the dotted bonds (there is a phosphorus atom where the three dotted lines meet) which enables cross-linking of the ribbons. Such complexity in allotropic forms is rivalled only by carbon and sulfur (see below).

The common α-forms of arsenic, antimony, and bismuth are ***isostructural*** and ***isomorphous*** with rhombohedral black phosphorus (Fig. 3.5(d)) and therefore similarly comprise puckered hexagonal sheets. Other allotropic forms of these three elements also exist. Thus, in addition to α-arsenic (usually called grey arsenic), elemental arsenic is found as yellow arsenic (comprising As_4 molecules analogous to P_4) and black arsenic which has the same structure as orthorhombic black phosphorus (Fig. 3.5(c)).

Other notable allotropes are the high-pressure metallic forms which for antimony is a hexagonal close-packed structure, and for bismuth is ζ-Bi in which each bismuth is surrounded by eight nearest neighbours in a body-centred cubic lattice.

Before leaving the Group 15 elements we should note an interesting trend in the interatomic distances in the isomorphous rhombohedral black phosphorus, α-As, α-Sb, and α-Bi series. As mentioned previously and shown in Fig. 3.5(d), each phosphorus atom in rhombohedral black phosphorus has three nearest neighbours arranged at the vertices of a trigonal pyramid. In addition, there are three other atoms at a greater distance *trans* to the three primary P–P bonds (in the adjacent sheet) such that the overall coordination is octahedral with three short and three long bonds. The observed trend is that, as the group is descended, the ratio of the long (r_2) to short (r_1) bonds decreases such that the bonds become more nearly equal in length; specifically, the values of r_2/r_1 for P, As, Sb, and Bi are 1.490, 1.240, 1.153, and 1.149 respectively. This is a definite trend from localized covalent bonding in a sheet structure towards a more delocalized, three-dimensional metallic bonding, and is consistent with the decreasing electronegativity of the elements as the group is descended. In the high-pressure cubic form of phosphorus, the effect of pressure is to shorten the long bonds to the point where $r_1 = r_2$.

3.6 Group 16

The most stable form of oxygen is dioxygen, O_2, which is a gas under ambient conditions and in which the two oxygen atoms are held together by a strong multiple bond (Fig. 3.6(a)). The only other form of oxygen which is stable under

Crystalline solids have long-range orientational and translational order, whereas amorphous solids do not.

In 2019, a new arsenic allotrope was reported which consists of a more complex layer structure comprising alternating units of the rhombohedral and orthorhombic structures.

Isostructural means that the gross structures are the same whereas isomorphous is a more precise term which means that the crystal structures are the same, i.e. the dimensions and symmetry of the unit cells are the same.

ambient conditions is ozone, O_3, (also a gas), which has a bent structure in which O–O multiple bonding is also important.

In contrast, sulfur does not exist as S_2 or S_3 molecules, except at high temperatures, but rather as S–S singly bonded species, the number of allotropes of which is quite large. We will not dwell on the details here except to say that besides the common form of sulfur which consists of S_8 rings (Fig. 3.6(b)), many other rings of different sizes have been characterized (S_n where n ranges from 6 to 20 though examples are not known for all values of n; even larger rings have been detected chromatographically in mixtures) together with polymeric, fibrous forms which contain helical chains of sulfur atoms. Both oxygen and sulfur are non-metallic, as expected from their electronegativity.

Selenium has properties consistent with its classification as a non-metal. Its allotropy, although much less extensive than that of sulfur, comprises modifications which contain Se_8 rings and a hexagonal form in which the selenium atoms form unbranched, helical chains (Fig. 3.6(c)). A structurally very complex vitreous form of selenium is also known in which rings of up to 1,000 atoms are thought to be present. Tellurium, in contrast, exists as only a single allotrope which is isostructural with hexagonal selenium. A similar trend in bond distances to that noted for the heavier Group 15 elements is apparent for the hexagonal forms of selenium and tellurium. Thus, the ratio of the short, covalent bond lengths to the longer interchain distances is closer to 1 for tellurium than for selenium, again indicating a trend to metallic character.

The final element in this group is polonium which is the least electronegative and most metallic in character. It has a structure in which each atom is surrounded by six nearest neighbours with a simple cubic unit cell—the only element in the periodic table for which this is found under ambient conditions.

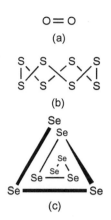

Fig. 3.6 (a) Dioxygen, O_2, (b) cyclooctasulfur, S_8, and (c) part of a chain of helical selenium, Se_x. Note that the helical forms of selenium and tellurium are chiral which is unusual for an element structure.

Polymorphism is also common in the elemental forms of sulfur. Thus, S_8 exists in one orthorhombic and two monoclinic modifications (in this context, modification is synonymous with polymorph).

3.7 Group 17

The halogens all exist as diatomic molecules under ambient conditions, F_2 and Cl_2 being gases, Br_2 a liquid and I_2 a solid. The low-temperature solid-state structures of Cl_2 and Br_2 are isomorphous with I_2 and all are layer structures, as illustrated for I_2 in Fig. 3.7. The solid-state structure of F_2, although different, is also layered (the structure of astatine is not known).

A number of further points are worth mentioning with regard to the solid-state structures of the halogens. Firstly, the ratio of the short, covalent bonds to the longer, intermolecular distances decreases as the group is descended in keeping with what we have seen in some of the structures in Groups 15 and 16. This is illustrated in Table 3.2 (distances in Å, the value for F_2 is in parentheses since it has a different crystal structure).

This observation is consistent with a trend towards increasing metallic character as a result of decreasing element electronegativity, and in this regard it is interesting to note that under high pressures, solid iodine does become metallic as evidenced by the onset of electrical conductivity. We can think of this as forcing the atoms together to such an extent that orbital overlap becomes sufficient

Table 3.2 Bond lengths (Å) for the halogens in the solid state

X	X–X	X⋯⋯X Intralayer	X⋯⋯X Interlayer	Smallest ratio X-X/X–X
F	1.49	3.24	2.84	(1.91)
Cl	1.98	3.32, 3.82	3.74	1.68
Br	2.27	3.31, 3.79	3.99	1.46
I	2.72	3.50, 3.97	4.27	1.29

We should also recognize that van der Waals or induced dipole-induced dipole interactions increase as the elements become more polarizable on descending the group which accounts for the trend in the physical states of the elements from gas to liquid to solid.

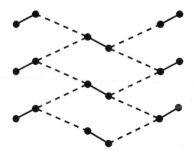

Fig. 3.7 A section of a layer in the solid-state structure of diiodine, I_2.

for extensive delocalization to occur which results in metallic bonding. A similar situation is thought to occur for dihydrogen, H_2, under very high pressures as well as for a number of metalloids including, as noted previously, phosphorus. We shall return to this in Section 3.9.

Another point to note is that the I–I bond distance in solid iodine (2.72 Å) is considerably longer than that in gaseous iodine (2.67 Å). The longer bond in the solid state is undoubtedly a result of the closeness of the intermolecular contacts—a feature we shall account for in Chapter 5 when dealing with the compounds of iodine (and other elements).

It is also noteworthy that the halogens become more intensely coloured as the group is descended: F_2 is essentially colourless, Cl_2 is pale green, Br_2 is deep red-brown, and I_2 is dark purple. We shall see that this trend of becoming more intensely coloured is also a feature of the compounds formed by these elements, and that an explanation has much in common with that for the trends in bond distances described in the previous paragraph.

3.8 Group 18

The Group 18 elements are all monatomic gases under ambient conditions and are entirely non-metallic. Whilst this certainly reflects their high electronegativity, the general lack of reactivity resulting from a filled valence shell is also an important factor. In the solid-state the Group 18 elements all adopt close-packed structures.

3.9 General features

An interesting insight into the relationships between many of the apparently very disparate structures of the elements in the p-block can be had by starting from the cubic diamond structure and seeing what changes arise as the number of valence electrons increases. For example, Fig. 3.8(a) shows a section of the structure of cubic diamond which, as we have seen, is also the structure adopted by silicon, germanium, and grey tin. Each carbon atom in cubic diamond forms four two-centre, two-electron C–C bonds, and there are no lone pairs of electrons. If we now move one group to the right (i.e. from Group 14 to Group 15), we now have one additional valence electron per atom or two electrons per pair of

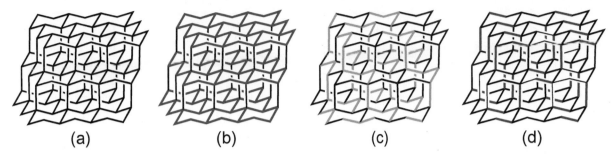

Fig. 3.8 Four diagrams showing the structure of cubic diamond. Diagram (a) represents cubic diamond itself and other Group 14 structures, while (b), (c), and (d) represent how the structures of the Group 15, 16, and 17 elements can be derived from the diamond structure as described in the text.

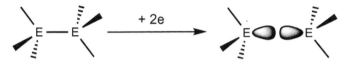

Fig. 3.9 The addition of two electrons to an E–E unit resulting in the breaking of one bond.

atoms. This leads to the breaking of one bond (out of four) for each atom such that this bond pair becomes two lone pairs, one on each atom as shown schematically in Fig. 3.9. The resulting atomic coordination is now trigonal pyramidal (three bond pairs and one lone pair) as opposed to tetrahedral. If the bonds broken are the vertical bonds in the cubic diamond structure shown in Fig. 3.8(a) (of which there is one per atom), the resulting structure now comprises parallel two-dimensional sheets of fused, puckered six-membered rings in the chair conformation as shown in red in Fig. 3.8(b). This is precisely the structure found in rhombohedral black phosphorus, α-As, α-Sb, and α-Bi.

The relationship of the layers in rhombohedral black phosphorus to each other is not exactly the same as the corresponding layers in cubic diamond. In rhombohedral black phosphorus the sheets have moved relative to each other to avoid close contacts between the lone pairs. Similar movements occur in the tellurium and diiodine structures with respect to the positions shown in Fig. 3.8 in order to minimize lone pair–lone pair contacts.

If we move to Group 16 there is now one more valence electron (i.e. two electrons per pair of atoms) which again leads to the breaking of one bond per atom which now results in two-coordination (two bond pairs and two lone pairs). If we do this in a particular way in the cubic diamond structure, this gives rise to a helical arrangement of atoms that is found in elemental tellurium and in one allotrope of sulfur and one of selenium (shown in green in Fig. 3.8(c)).

For the elemental structures described here, the number of bonds formed by any particular element is given by the so-called 8 – N rule, where N is the number of valence electrons.

Finally, the addition of a further electron per atom takes us to Group 17 and we break another bond so that each atom is now one-coordinate with three lone pairs. One way of doing this results in the structure adopted by dichlorine, dibromine, and diiodine in the solid state and is represented in blue in Fig. 3.8(d).

Before leaving the cubic diamond structure, we can also note that if we proceed as far as the rhombohedral black phosphorus structure shown in Fig. 3.8(b) and then flatten the puckered hexagonal sheets into planar hexagonal sheets, we arrive at the rhombohedral graphite structure referred to earlier. The structural change from diamond and graphite in this case is not driven by increasing the number of valence electrons; rather, it is an alternative structure for four valence

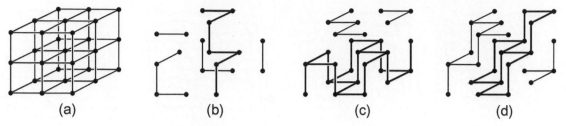

Fig. 3.10 Four diagrams showing (a) a simple cubic lattice and how the structures of (b) tellurium, (c) orthorhombic black phosphorus, and (d) rhombohedral black phosphorus can be derived from this lattice.

electrons, the two structures, in the case of carbon, being similar in energy. This is a good illustration of the importance of the strength of π-bonding in first-row element structures (for reasons we shall see in the next chapter and which will account for why there is no counterpart to the graphite structure in the elemental forms of silicon and germanium).

An alternative approach to understanding the relationships between the structures of the p-block elements is in terms of Peierls distortions (introduced in Section 3.1) away from a symmetric reference structure. Thus, if we take a simple cubic lattice as shown in Fig. 3.10(a) (which represents the structure of α-polonium and a high-pressure allotrope of phosphorus), we can arrive at the helical sulfur, selenium, and tellurium structure by removing four lines per lattice point, as illustrated in Fig. 3.10(b); one particular unit is shown in bold to illustrate the helical structure. If, however, we start from the simple cubic lattice and remove three lines per lattice point, we can, according to the precise arrangement of lines removed, derive either the structure of orthorhombic black phosphorus (Fig. 3.10(c)) or the structure of rhombohedral black phosphorus plus α-As, Sb, and Bi (Fig. 3.10(d)); bold is again used to highlight the important structural motif, and Figs. 3.10(b)–(d) should be compared with Fig 3.5(c) and (d) and with Fig. 3.8(b) and (c). A full treatment of this approach based on Peierls distortions is beyond the scope of this book (more information can be found in the many papers of J. K. Burdett on this subject), but it is possible to derive the structures of most of the elemental forms of the p-block elements along these lines (boron is an exception) including graphite and the white tin and gallium structures.

In concluding this section we can see that a seemingly very diverse range of element structures can be rationalized in a way that reveals them to have a number of structural similarities. While it is true that these may seem to be somewhat *post hoc* rationalizations with little predictive power (there are many ways to remove bonds and lines between lattice points), the similarities and relationships these analyses highlight offer useful insights.

These approaches—particularly that associated with a Peierls distortion—also offer an insight into the effect of increasing pressure. We have noted the cubic form of phosphorus which can be prepared at high pressure, and the structure of this allotrope is as shown in Fig. 3.10(a). Whether starting from the orthorhombic or rhombohedral forms of phosphorus, it is clear that pushing the

Removing lines is akin to lengthening or breaking bonds.

atoms closer together (which is what increasing the pressure does) provides an easy pathway from either of those structures to the simple cubic form; you just make the longer bonds shorter until they are the same length as the short ones. One of the consequences of this structural change is that the bonding becomes more delocalized which results in a change from non-metallic to metallic properties such as insulating to conducting. While structurally different, increasing pressure results in a conducting form of iodine and, although it is not yet experimentally confirmed, it is predicted that under very high pressures, hydrogen becomes metallic. If we again consider the rhombohedral black phosphorus to cubic phosphorus transition where long bonds become shorter (i.e. the long bond to short bond ratio becomes smaller), this essentially mimics the trend seen when descending the group moving from phosphorus to bismuth (and indeed, sulfur to tellurium).

3.10 Binding energies

A final point of interest whilst dealing with the elements themselves concerns the so-called binding energies (or atomization energies) which relate to the magnitude of the interatomic forces which bind the elements together.

For example, in a diatomic molecule, the binding energy (more often called the bond energy) is defined according to Eqn. 3.1 whereas for a macromolecular solid, Eqn. 3.2 is appropriate.

$$X_2(s) \rightarrow 2X(g) \tag{3.1}$$

$$X_n(s) \rightarrow nX(g) \tag{3.2}$$

A graphical representation of the binding energies for the elements of the first two rows of the s- and p-block is shown in Fig. 3.11 from which it is apparent that the energies peak in Group 14 and fall to zero for the Group 18 elements. The nature of this graph simply reflects the number of valence electrons available for bonding. The trend from lithium to carbon in the first row is a consequence of the number of valence electrons increasing from one to four. For nitrogen, although there are five valence electrons, only three are used in bonding (the remaining two are a lone pair) and so the binding energy is similar to that of boron (note that this is true even though elemental boron and nitrogen have quite different structures and very different melting and boiling points). Similarly, sulfur with six valence electrons uses only two in element–element bonding and therefore has a similar binding energy to beryllium. In general, the reactivity of an element will increase as its binding energy decreases although the noble gases in Group 18 are an obvious exception.

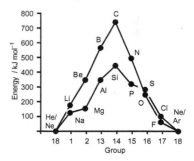

Fig. 3.11 The binding energies for the elements of the first two rows.

Strictly speaking, the binding energies of the Group 18 elements are not zero since the atoms are attracted by van der Waals forces, but these are small in comparison with the binding energies associated with shared electrons, i.e. covalent bonding.

4 General features of s- and p-block element compounds

Before considering the various classes of s- and p-block element compounds and their properties, which we will do in Chapter 5, it will be useful to look at some general aspects, such as trends in common oxidation states, and the effect of atomic size on coordination numbers (Sections 4.2 and 4.3) which, together with properties such as electronegativity described in Chapter 2, will form the basis of our understanding of the structure, properties, and reactivity of s- and p-block element compounds. In addition, we will look at trends in bond strengths (Section 4.4), and finally at some general observations regarding compound types using the so-called van Arkel–Ketelaar triangle (Section 4.5). Prior to all of this, however, we shall consider some important concepts, particularly those of oxidation state and valence (Section 4.1).

4.1 Definitions of oxidation state and valence

We shall start by reviewing the terms 'oxidation state' and 'valence' and how they are used. We will also look at some related terms, and one of the reasons for spending time on this matter is that many of these terms or concepts are often used interchangeably and sometimes wrongly in a way that can lead to confusion. References are given in the margin to some of the literature on this topic which also provide a useful historical background.

Let us start with NF_3 as an example (Fig. 4.1(a)). In assigning an **oxidation state** (sometimes called **oxidation number**) we partition the elements according to their electronegativities. Thus, since fluorine is more electronegative than nitrogen, this is formally a compound of nitrogen(III) with fluorine present as fluorine(–I). We could write this as N^{3+} and F^- (i.e. fluoride), but it is better to adhere to the Roman numeral designation, since we must recognize that oxidation states are a formalism (albeit a very useful one as we shall see), and there is no sense in which NF_3 should be thought of as an ionic compound with N^{3+} and F^- ions (because it is not!); N(III) and F(–I) are therefore examples of the designations we shall mostly use when describing oxidation states, although sometimes oxidation states may be encountered in a form such as N^{III}. Now, what about **valence** (sometimes called **valency**)? In NF_3, nitrogen is using three

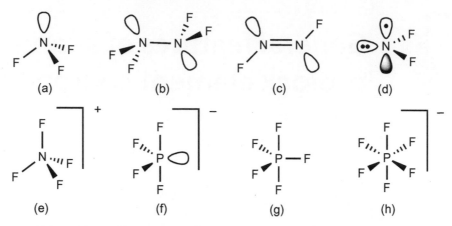

Fig. 4.1 The structures of (a) NF_3, (b) N_2F_4, (c) N_2F_2, (d) NF_2, (e) $[NF_4]^+$, (f) $[PF_4]^-$, (g) PF_5, and (h) $[PF_6]^-$. Lone pairs are also shown, and, in the case of NF_2, all the non-bonding electrons on the nitrogen are indicated.

Probably the most definitive text on oxidation states is the IUPAC Technical Report on oxidation states entitled *Toward a Comprehensive Definition of Oxidation State*, authored by Karen, McArdle and Takats (*Pure Appl. Chem.* 2014, **86**, 1017). The definition of oxidation state given is: *the oxidation state of a bonded atom equals its charge after ionic approximation*, and this is the approach adopted here.

of its valence electrons (out of a possible five) to form bonds to fluorine and is therefore trivalent. Strictly, it is the number of valence electrons being used that defines valence or the valence number which is three in this example. The fact that NF_3 has three N–F bonds is not sufficient to classify the nitrogen as trivalent, but we can classify it as having a **covalence (covalency)** of three based on the fact that there are three covalent bonds present. This distinction will be important in a later example.

So much for NF_3; what about N_2F_4, the structure of which is shown in Fig. 4.1(b)? If we partition again according to electronegativity, we have four fluorides (F⁻), leaving an N_2^{4+} unit which is therefore N^{2+} for each nitrogen. This is therefore a compound of nitrogen(II), but each nitrogen still uses three valence electrons and still forms three covalent bonds (two N–F and one N–N) so retains a valence of three and a covalence of three. Likewise, N_2F_2 (Fig. 4.1(c)) is a compound of N(I) but also with a valence of three and covalence of three, since the two nitrogens are joined by a double bond.

Before leaving nitrogen fluorides, let us consider two others. First, N_2F_4 readily dissociates into NF_2 radicals (Fig. 4.1(d)). According to the oxidation state formalism this is still a compound of N(II), but it is divalent, since the nitrogen is using only two of its valence electrons in bonding; with two N–F bonds, NF_2 also has a covalence of two. This leads us to the recognition of an important point; N_2F_4 and NF_2 are both formally N(II), but the former is trivalent and the latter divalent. It would be correct to refer to NF_2 as sub- or low-valent ('sub' and 'low' being relative to a more usual higher valence), but it would not be correct to describe N_2F_4 as sub- or low-valent. Such a distinction is not always appreciated, however, and reflects an erroneous conflation of the terms 'oxidation state' and 'valence'.

The point about valence and covalence in ionic species is further illustrated by ions such as $[NF_2]^+$ and $[NF_2]^-$, which each have a covalency of two, but the former is trivalent whereas the latter is monovalent. The oxidation states are N(III) and N(I) respectively.

The final nitrogen fluoride we shall look at is the ion $[NF_4]^+$ (Fig. 4.1(e)). According to the oxidation state formalism, this is a compound of N(V) (remove four fluorines as F⁻ from a species with a single positive charge). Nitrogen is using all five of its valence electrons, so it is also pentavalent, but it has a covalence of four since it contains four N–F bonds. Thus, while valence and covalence may be the same in neutral compounds, they may not be in certain ionic species.

Now let us look at some corresponding phosphorus fluorides (Fig. 4.1(f)–(h)). For those which are heavier congeners of the nitrogen fluorides we have just examined, everything in terms of oxidation state (fluorine is more electronegative than phosphorus) etc. is the same. However, phosphorus also forms some species that have no counterpart in nitrogen chemistry, one such being the anion $[PF_4]^-$. This is a compound of P(III) (four fluorines but note the negative charge), and has a phosphorus valence of three and an apparent covalence of four (four P–F bonds). We should also consider two other compounds: PF_5 and $[PF_6]^-$. It is straightforward to determine that these are both compounds of P(V) and both contain pentavalent phosphorus, but what about their covalence? If we describe these compounds as having two-centre, two-electron bonds, they clearly have covalencies of five and six respectively, but in using the term covalence we need not restrict ourselves to two-centre, two-electron bonds. Whatever the precise nature of the bonding implied by the line drawn between the atoms, some covalent bonding exists, and the sum of these (i.e. the number of bonds as drawn) can be used to define the covalence number; in the case of $[PF_4]^-$, PF_5, and $[PF_6]^-$, this is four, five, and six respectively.

If we do adopt a bonding model in which all the bonds are indeed two-centre, two-electron bonds, both PF_5 and $[PF_6]^-$ seemingly violate the octet rule in that they have ten and twelve electrons around the phosphorus centre respectively ($[PF_4]^-$ also has ten electrons). Compounds which violate the octet rule are often (since the 1960s) termed **hypervalent**. As we will see later, however, we can describe the bonding in $[PF_4]^-$, PF_5, and $[PF_6]^-$, and other so-called hypervalent compounds, in terms of a multi-centre bonding model such that the octet rule is not violated. In so doing, we will arrive at a more satisfactory description of the bonding and will also see that the term hypervalent is not only unnecessary but rather misleading. To briefly pre-empt this later discussion, the anion $[SiF_5]^-$ has been described as hypervalent (it is isoelectronic with PF_5), but based on the methodology outlined previously, this species contains Si(IV) and is tetravalent; what is 'hyper' about that? When people talk about hypervalent, what they generally mean is hypercovalent (or perhaps hypercoordinate) in circumstances where the octet rule is (apparently) violated, but we shall return to this in Chapter 6.

With the exception of the term hypervalent (or hypercovalent), all of the other terms we have looked at so far are pretty clear and, as defined, largely unambiguous. In part, though, this is because of the examples chosen to illustrate them. Matters can be a little more complicated, as we shall see from considering $POCl_3$.

The compound $POCl_3$ is often drawn as shown in Fig. 4.2(c), but can just as well be represented in any of the forms (a)–(d). We will not dwell on which is the best representation of the bonding or electronic structure, but we can state that (b)–(d) all have P(V) and pentavalent phosphorus. Covalencies would seem to range from four to six, however. What about Fig. 4.2(a)? With a dative bond from phosphorus to oxygen this might be described as P(III) and trivalent phosphorus with a covalence of three or four, depending on whether we include the dative bond or not. However, since the phosphorus is using all of its valence electrons

There is a notation X-E-Y sometimes used for compounds such as $[PF_4]^-$, PF_5, and $[PF_6]^-$, where E is the central element, X is the total number of valence electrons associated with E, and Y is the number of bonds. The three phosphorus examples would therefore be denoted as 10-P-4, 10-P-5, and 12-P-6 species respectively.

The term 'hyper' is defined in the *Oxford English Dictionary* as meaning 'over', 'above', or 'beyond'. If it is to have any value in this context it is not really useful to consider N(V) or P(V) as hypervalent except in the limited sense that (V) is greater than (III). As noted in the main text, if the term hypervalent is appropriate at all, it is in relation to compounds that seemingly violate the octet rule.

A notation sometimes employed is of the form σ^x, λ^y, where σ is the coordination number and λ is the number of bonds (i.e. a covalence), but specifically the number of bonds in representations such as Fig. 4.2(c). $POCl_3$ would therefore be described as σ^4, λ^5. The coordination number is simply the number of atoms to which the central atom is bonded.

(albeit with two forming a dative bond), it should be considered pentavalent (this might seem obvious, but in coordination complexes containing phosphine ligands, the phosphorus is often considered as trivalent). In terms of hypervalence (bearing in mind what has been said about this term above), neither (a) nor (b) would (or should) be considered as such, but (c) and (d) would. The lesson to be learned, then, is that in anything other than the simplest of cases, ambiguities may abound according to how the bonding is described, except in the case of the terms **valence** and **oxidation state** which have precise definitions.

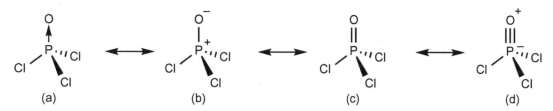

Fig. 4.2 Representations of the bonding in $POCl_3$ (a)–(d).

We could leave our discussion of oxidation state and valence here, but it would be remiss not to mention the work of Green and Parkin who have developed the so-called Covalent Bond Classification (CBC) methodology specifically for the teaching of these topics and to avoid issues that some perceive with the use of oxidation states. The CBC is sometimes referred to as the MLX method where M is a central atom (often but not exclusively a transition metal which is where the method finds most application), L is a group which donates two electrons to the central atom (a ligand), and X is a group which provides one electron to the central atom. There are also Z groups which are electron pair acceptors with respect to M. One of the principal reasons (out of several) claimed as an advantage of the CBC method can be appreciated if we look at a series of carbon compounds such as CH_4, CH_3F, CH_2F_2, CHF_3, and CF_4. Since the ordering of electronegativity is H<C<F, the oxidation state of the carbon in these compounds is, respectively, C(–IV), C(–II), C(0), C(II), and C(IV). All, however, are tetravalent and have a covalency of four. This wide range of oxidation states is seen as a little perverse and, according to the authors of the CBC, reveals the shortcomings of the oxidation state formalism whereas within the CBC method, all of these compounds would be classified as MX_4 species. This is not the place to go into detail about the pros and cons of the various methodologies described above, and the reader is referred to the references in the margin for a fuller description of the CBC approach. Suffice to say that the classification of molecules in terms of oxidation state and valence is replaced in the CBC with the terms **electron number** and **valence number** (together with the **ligand bond number**) which are defined in the margin.

More detail on many of the issues discussed here, especially the CBC method, can be found in M. L. H. Green and G. Parkin, *J. Chem. Ed.*, 2014, **91**, 807, and G. Parkin, *J. Chem. Ed.*, 2006, **83**, 791. Another useful article, and one which makes the case for the term 'covalence', is D. W. Smith, *J. Chem. Ed.*, 2005, **82**, 1202.

For a molecule $ML_lX_xZ_z$, the following definitions apply: The **electron number (EN)** = $m + 2l + x$, where m is the number of valence electrons on the neutral M. The **valence number (VN)** = $x + 2z$. The number of non-bonding valence electrons (n) = $m - x - 2z = m - VN$. The **ligand bond number (LBN)** = $l + x + z$ (sometimes, but not always, the coordination number). VN is the same as the valence described in the text.

Two more points are worth making in relation to the CBC method. The neutral nitrogen and phosphorus compounds mentioned previously are easily accommodated in the CBC method, e.g. MX_3 and MX_5 for NF_3 and PF_5 respectively. Ions can be represented in two ways, however. Thus NF_4^+ (we shall dispense with the square brackets) can be described as MX_4^+, but it can also be thought of as an M with three X groups and one Z group such that the so-called **equivalent neutral** *class* is MX_3Z. Similarly, PF_4^- can be written as MX_4^- or, in terms of its equivalent neutral class, as MX_3L; if you like, phosphorus bonded to three fluorines and one fluoride as opposed to three fluorines and an F^+ in the previous example. I will leave it to the reader to decide which you prefer.

Returning briefly to organic compounds, if the case is made (see Calzaferri *J. Chem. Ed.*, 1999, **76**, 362) to treat hydrogen as neutral in terms of assigning an oxidation state, the oxidation state of carbon in CH_4 would be C(0), thereby reducing the spread in oxidation states for the carbon compounds mentioned previously in relation to the CBC approach. In so doing, the resulting **oxidation level** of carbon (as it is often called) for several organic compound classes (for the most part, functional groups) can be assigned as shown below which simply reflects the heteroatoms bonded to the carbon. This oxidation level approach is particularly useful in deciding whether an oxidizing agent or a reducing agent (or neither) is required for any particular transformation (which is why it is used), and this is a strength of the oxidation state formalism generally.

Heteroatoms are (usually) more electronegative than carbon, but note that C(O) or C=O is equivalent to $C(OH)_2$ in the same way that it would be in calculating a formal oxidation state. In fact, oxidation level tends to be used in a qualitative sense (chemists might talk about the ketone oxidation level) rather than using the numerical scale shown here, but the term is perfectly amenable to the quantification illustrated.

Oxidation Level 4: CF_4, $C(O)(OR)_2$, CO_2

Oxidation Level 3: $RC(O)X$, X = OH, OR, NR_2, Cl

Oxidation Level 2: RC(O)R/H

Oxidation Level 1: Alkyl halides, alcohols, amines

Oxidation Level 0: CH_4, alkanes

Finally, let us briefly consider CaF_2. This is an ionic compound with a typical ionic structure which we shall look at in Chapter 5. The oxidation state of calcium is clearly Ca(II) (the designation Ca^{2+} would be acceptable here), but it makes no sense to talk about its covalence since it is an ionic compound. The calcium is divalent in the sense that it is using its two valence electrons.

In solid-state chemistry, the concept of **bond-valence** (V_A) is often employed. Thus, the valence V_A of A in a compound AX_n is distributed over all the A–X bonds according to the valence sum rule $\Sigma s_{AX} = V_A$, where s_{AX} is the bond-valence of a particular bond and is correlated with the length of that bond (D_{AX}) according to the expression $s_{AX} = \exp[(r_0 - D_{AX})/b]$, where r_0 and b are empirically determined parameters. In general, V_A is also equal to the formal oxidation state of A.

4.2 Common oxidation states

General points

Let us now turn to some general features regarding the common oxidation states which are found for compounds of the s- and p-block elements (ignoring therefore the oxidation state of 0 for the elements themselves). We shall focus, for the most part, on oxidation state rather than valence or covalence unless we specifically need to consider the valence terms, which, on occasion, it will be useful to do. We shall also restrict ourselves mostly to mononuclear compounds.

Probably the most notable feature is that for elements with more than one available oxidation state, the formal numbers of the common oxidation states usually differ by two units as indicated in Table 4.1. The oxidation states shown do not constitute an exhaustive list, and there are some important exceptions such as NO and NO_2 in Group 15 which are formally nitrogen(II) and (IV) respectively and which we shall look at below.

The key aspect to note is that elements in their common oxidation states listed in Table 4.1 contain no unpaired electrons whereas the two nitrogen oxides just

Table 4.1 Common oxidation states for compounds of the s- and p-block elements

Group	1	2	13	14	15	16	17	18
Oxidation states	1	2	3, 1	4, 2	5, 3, –3	6, 4, 2, –2	7, 5, 3, 1, –1	6, 4, 2

It is not that compounds with E–E bonds and therefore oxidation states which differ from those listed in Table 4.1 are especially uncommon, certainly in p-block chemistry, but, as noted previously, we shall focus principally on mononuclear compounds. Nevertheless, it is important to be aware of the precise definitions of oxidation state and valence set out in Section 4.1.

Delocalization in a π^*-orbital is also the case in the chlorine(IV) species ClO_2, which is another example of a p-block compound containing an unpaired electron.

mentioned are both radicals, i.e. they *do* contain unpaired electrons. The reason that radicals are not all that common for mononuclear compounds in s- and p-block chemistry is that the unpaired electrons would generally reside in p- or sp^n-type hybrid orbitals which have fairly good directional properties. Overlap between these orbitals therefore tends to be favourable, and reasonably strong bonds are formed such that there is a marked tendency for such radicals to dimerize to form species with element–element bonds unless the unpaired electrons are delocalized or the resulting E–E bond is particularly weak for some reason.

Delocalization is the situation in NO where the unpaired electron resides in a N–O π^*-orbital. A further consideration in this case is that although NO can dimerize to give N_2O_2 (which does exist in the solid state), dimerization results in no net change in aggregate bond order. Thus, the bond order in NO is 2.5 since the unpaired electron resides in a π^*-orbital, i.e. exactly half that of N_2O_2, which is five in total (O=N=N=O, bent at each nitrogen). There is, therefore, no net increase in bonding upon dimerization, at least to a first approximation, as would more usually be the case in the dimerization of radicals.

In the case of NO_2, the weak N–N bond in the dimer N_2O_4 is the reason why appreciable dissociation into NO_2 radicals occurs. The N–N bond in N_2O_2 is also weak, as is the N–N bond in N_2F_4 (as noted in the previous section); we shall see why N–N bonds are weak when we address bond energies in Section 4.4.

We should note at this point that the rarity of radicals (compounds, specifically mononuclear compounds, with unpaired electrons) in s- and p-block chemistry is in stark contrast to the situation found in compounds of the d- and f-block elements, particularly for the 3d and 4f elements. Here, compounds or complexes with unpaired electrons are frequently encountered, and common oxidation states often differ by only one unit, e.g. Fe(II) and Fe(III). This can be traced to the fact that the valence electrons in d- and f-block complexes are generally in d and f orbitals respectively, which, we might argue, are more diffuse and less directional than corresponding s-p hybrids. The tendency for compounds with unpaired electrons to dimerize with the formation of element–element bonds is therefore not so pronounced for this reason. Other factors are also important, however, such as the often relatively small radial extension of d and f orbitals and the fact that in d- and f-block element complexes, the element centres are surrounded by ligands.

Let us now look in more detail at the common oxidation states encountered in the s- and p-block groups.

In **Group 1** and **Group 2**, the oxidation states are +1 and +2 (we will use Arabic numerals as well as Roman numerals in these subsections) respectively resulting from loss of the valence s electrons, and any higher oxidation states are precluded by the prohibitively large energies required to remove electrons from the next quantum shell down. The reason why compounds of the Group 2 elements in the +1 oxidation state are not generally observed is less obvious, but thermodynamic considerations reveal that disproportionation reactions (Eqn. 4.1) for simple salts EX (X = halide, for example) are strongly exothermic being driven by the large lattice energy of the resulting EX_2 salt. Some compounds of the Group 2 elements are known in the formal +1 oxidation state, but these contain element–element bonds and are therefore divalent; a magnesium example is shown below.

$$2 \, E(I)X \rightarrow E(0) + E(II)X_2 \qquad\qquad (4.1)$$

We should also note that most of the Group 1 elements form compounds (usually salts) containing the element in the –1 oxidation state, e.g. Na(–I) in $[Na(\text{cryptand}[2,2,2])]^+Na^-$ where cryptand[2,2,2] is a nitrogen-containing ligand which encapsulates the Na^+ in order to keep it separated from the Na^- anion.

If we consider **Group 13**, the common oxidation states are +3 and +1. The former arises by formal loss of the three valence electrons, $ns^2\,np^1$, which gives a +3 centre with a noble gas core. (Remember that we are dealing with a formalism here; in TlF_3 it is reasonable to view this compound as predominantly ionic with a Tl^{3+} cation and three fluoride anions, but for the much more covalent InI_3 the charges are purely formal.)

The +3 state is observed for all elements in this group, but the +1 state, which results from removal of the single valence p electron, is common only for the heavier elements, particularly so for thallium. In fact, the +1 oxidation state is probably more important than the +3 state in the chemistry of thallium, and we shall see that oxidation states of two less than the group maximum are commonly observed for the 6p elements. This is usually referred to as the ***inert pair effect***, and we shall consider the origin of this effect below.

In **Group 14**, the common oxidation states are +4 and +2 (bearing in mind the issue around oxidation states in carbon compounds noted in Section 4.1, so we shall also start to use valence and/or covalence interchangeably with oxidation state where it is correct to do so) with the latter being most common for tin and lead. The inert pair effect is particularly important for lead (as it was for thallium and for the same reasons, see below) such that Pb(II) is the dominant oxidation state in lead chemistry, although we must be a little careful of generalizations of this type. For example, it is found that whilst lead(II) chloride, $PbCl_2$, is indeed more stable than lead(IV) chloride, $PbCl_4$, in line with the trends described

For the 6p elements thallium, lead, and bismuth, the point about lower oxidation states is certainly true, and it would seem to be for polonium as well; the only examples of Po(VI) are PoF_6 and PoO_3. For astatine and radon, experimental data are rather scarce due to the highly radioactive nature of these elements.

previously (see also Chapter 5), the opposite is true for the organometallic alkyl compounds; e.g. tetraethyl lead, $PbEt_4$, is stable, whereas diethyl lead, $PbEt_2$, is not (a similar situation is found for the chlorides and alkyls of thallium(III) *vs* thallium(I)). We shall return to this matter in Section 4.4.

In **Group 15**, the common oxidation states are +5, +3, and −3 or, more simply, compounds are generally pentavalent or trivalent. The inert pair effect can be (and often is) said to have some influence on the importance of bismuth(III) over bismuth(V), but with oxidation states as high as +5 there are other factors which are important in determining the relative stability of possible oxidation states. Thus, high positive oxidation states tend to be good oxidizing agents since the high formal charge centres are good at attracting electrons and thereby being reduced. As such, high oxidation state centres (let us say +5 and above) are generally only found in association with fluorine or oxygen since these very electronegative elements are themselves difficult to oxidize (This feature of high element oxidation states being found in association with fluorine and oxygen is not restricted to the p-block, it is common in the d- and f-blocks as well; the low oxidation states of the s-block elements require no special stabilizing conditions.) As an example, PF_5 is a known compound, whereas PI_5 is not.

In **Group 16** we find that the +6 oxidation state is, not surprisingly, found most often in conjunction with fluorine and oxygen although, as in the case of lead(IV), this is a generalization which can be taken a little too far. For example, tellurium hexamethyl, $TeMe_6$, has been synthesized and found to be relatively stable (and in Group 15, $BiMe_5$ is also known along with several bismuth penta-aryl compounds, $BiAr_5$). The other stable states are the +4 and the +2, although in the case of oxygen, oxidation states greater than +2 are not observed. This is a consequence of the high electronegativity of oxygen and, strictly speaking, it will only be found in a positive oxidation state with elements of greater electronegativity which means fluorine; OF_2 is formally O(II). In H_2O, oxygen is more electronegative than hydrogen and is therefore formally oxygen(−II). In both OF_2 and H_2O the oxygen is, of course, divalent (and dicovalent), and this is characteristic of the element. Sulfur is also often found as sulfur(−II), though in the presence of oxygen or fluorine, compounds of S(II), S(IV), and S(VI) are well characterized.

In **Group 17**, the elements exhibit the widest range of oxidation states in the p-block. Fluorine, being the most electronegative element in the periodic table, is never found in formal positive oxidation states and exists as F(−I) in all of its compounds. The heavier halogens, however, are found in oxidation states −1, +1, +3, +5, and +7 although the higher oxidation states are usually powerful oxidizing agents and are only found in association with very electronegative atoms or groups such as oxygen or fluorine.

In **Group 18** the most common oxidation state is zero, i.e. the atomic elements themselves, and this is the state found exclusively for the elements He, Ne, and Ar. Krypton is known in the +2 oxidation state as KrF_2, but only for xenon is the chemistry extensive, with oxidation states of +2, +4, and +6 being observed (+8 is known in a few examples) albeit very largely in the presence of electronegative elements or groups. The reason for positive oxidation states being more common for the heavier elements simply reflects their lower ionization energies.

In considering the differences between PF_5 and PI_5, it may well be correct to say that PI_5 does not exist because P(V) would oxidize iodide (and we would end up with PI_3 and I_2), but we must also take size effects into consideration as well; iodide is large, and it is probably not possible to place five iodines around a small P(V) centre. Salts containing the $[PI_4]^+$ cation are known, however.

In the case of oxygen, hydrogen peroxide, H_2O_2, is formally oxygen(I) (with divalent oxygen) and the ionic species O_2^+, O_2^- and O_2^{2-} derived from O_2 all have atypical valences and oxidation states. Note also that in the hydronium ion $[H_3O]^+$, the oxygen is formally tetravalent since it is using four of its valence electrons in bonding.

This text deals mostly with elements or compounds at or near room temperature and atmospheric pressure. There is, however, an increasing knowledge of chemistry at very high pressures, and under these conditions, compounds involving even elements such as helium have been characterized, and recent calculations indicate that caesium may form a number of higher fluorides under pressure such as CsF_3, a compound of Cs(III).

The potentially extensive chemistry of radon is limited by the handling problems associated with its radioactivity.

Having considered some general points for the various groups in the s- and p-block, let us turn to some more specific matters.

The inert pair effect

We highlighted the so-called inert pair effect specifically in relation to the stability of Tl(I) and Pb(II) (but also for Bi(III) and Po(IV)). A standard and often quoted explanation for the inert pair effect is that the relatively large effective nuclear charge of thallium and lead resulting from the presence of the 4f and 5d elements, coupled with relativistic effects, leads to a substantial stabilization of the 6s pair relative to the 6p electrons, and it is this factor which gives rise to the term 'inert pair' itself, i.e. an inert 6s pair. However, this is only part of the explanation and, indeed, the term must be seen as something of a misnomer in this regard. Although the 6s pair is stabilized by these effects, it can hardly be considered to be too low in energy (i.e. core-like) in view of the relevant ionization energies, which are comparable with those of the lighter elements. This is illustrated in Table 4.2 in which values for the sum of the first three ionization energies (Σ IE 1–3) for the Group 13 elements are presented.

Table 4.2 Values for the sums of the first three ionization energies for the Group 13 elements (kJ mol⁻¹)

Element	B	Al	Ga	In	Tl
Σ IE 1–3	6888	5140	5521	5084	5438

All the values in Table 4.2 are of a comparable magnitude, and we therefore cannot argue that the 6s pair of thallium is unavailable for bonding because it is too low in energy (we could make the same argument based on orbital energies shown in Fig. 2.6 in Chapter 2). If this argument were appropriate at all, we would expect that, of all the Group 13 elements, boron would exhibit an inert pair effect since the s pair ionization energies are significantly larger for this element—a factor which is also apparent from the graph of orbital energies shown in Fig. 2.6, and one which would be particularly important for the elements further to the right in the 2p row.

A consideration of the experimental electron promotion energies required to achieve the trivalent and tetravalent states (the so-called valence states where all electrons are unpaired) of Tl and Pb respectively, shown in Fig. 2.7 in Chapter 2, is, perhaps, more appropriate, especially when dealing with covalent compounds, and here we should note that the values for Tl and Pb are the largest in their groups. However, whilst this will clearly be important, they are not that much greater than the values for their lighter congeners and cannot, therefore, constitute a complete explanation.

In fact, the real origin of the inert pair effect is probably a combination of two factors. The high promotion energy involved in achieving the valence state is one factor, but an equally important point is the weak bonds formed as a result of the large size

Note that the trend in ionization energies in Table 4.2 mirrors that found for the electronegativities of these elements, although this is not surprising since the Mulliken electronegativity scale, for example, actually uses ionization energies to calculate electronegativities. Thus, the electronegativity of gallium is higher than aluminium, and that of thallium is higher than indium.

As noted earlier, the explanation for the high promotion energy of Tl and Pb is due to spin–orbit coupling effects rather than any difference in the orbital energies. Spin–orbit coupling increases with n (the principal quantum number) and Z^*, this latter factor accounting for the irregularity of the values in Fig. 2.7. Increased spin–orbit coupling with increasing n is a consequence of relativistic effects, as noted in Chapter 2 (see Dasent (1965)).

of the heavier atom. Thus, for boron, the energy required to involve the 2s pair in bonding (the promotion energy; the ground state is $2s^2\,2p^1$ whereas the valence state is $2s^1\,2p^2$) is more than offset by the formation of three strong covalent bonds, and boron is therefore more likely to occur with an oxidation state of +3 rather than +1, or, more generally, it will tend to be trivalent rather than monovalent. In the case of thallium, the opposite tends to be true, and the formation of two extra, much weaker bonds associated with the +3 oxidation state, or trivalency, is not enough to compensate for the considerable promotion energy. Exactly the same argument can be advanced for carbon vs lead, but there are some interesting caveats in lead chemistry to which we shall return in Section 4.4.

We should also consider the significant difference in the radial extension or size of the 6s and the 6p orbitals as mentioned in Chapter 1. The importance in this context is that s-p hybridization will be less efficient leading to weaker covalent bonding in situations where s orbital participation is required (such as trivalent thallium or tetravalent lead). We shall meet this idea again and in more detail in later sections.

Finally, a very brief mention of the predicted chemistry of the newly identified and named 7p elements. The relativistic non-degeneracy of the p orbitals resulting in a $p_{1/2}$ orbital that is lower in energy than two $p_{3/2}$ orbitals described in Chapter 2 is expected to be especially significant for these elements. Thus, the elements Flerovium (Fl, Group 14) and Moscovium (Mc, Group 15) are predicted to exhibit a double inert pair effect in which a 7s (strictly a $7s_{1/2}$) pair and $7p_{1/2}$ pair remain non-bonding which would result in the oxidation states Fl(0) and Mc(I) being dominant in the chemistry of these two elements. Fl(0) might therefore have some similarities with Hg(0).

Elements of the 4p row

A feature of the elements of the 4p row, i.e. Ga–Kr, which warrants an explanation is their reluctance to exhibit or achieve their group maximum oxidation state, examples of which we shall meet in later chapters. This bears some similarity to the matter described in the previous section and could be considered as another example of an inert pair effect, although it is not usually referred to as such. Any effect here is not so apparent for the respective +3 and +4 oxidation states of gallium and germanium, but compounds of arsenic(V), selenium(VI), and bromine(VII) are much less common than compounds of phosphorus and antimony(V), sulfur and tellurium(VI), and chlorine and iodine(VII) respectively. Moreover, compounds of the 4p elements which do exhibit the group maximum oxidation state are generally more powerful oxidizing agents than their lighter and heavier congeners.

The reason for the high oxidation states of the 4p elements being uncommon is likely to be similar to that encountered for the related situation with the 6p elements. Thus, the increased effective nuclear charge associated with the presence of the preceding 3d row will make it more difficult to involve all the valence electrons in bonding, and any extra bonds formed may not be sufficiently strong to compensate. With gallium and germanium, the effect is not so dramatic in that

The inert pair effect should favour the zero oxidation state for mercury, the element one to the left of thallium in the periodic table which accounts, at least in part, for why mercury metal is rather unreactive and why the element is a liquid at room temperature. Relativistic effects are crucial to understanding many aspects of mercury chemistry such as strong Hg–Hg bonding in the Hg_2^{2+} ion, i.e. Hg(I).

Another consequence of the inert pair effect concerns the ubiquitous lead-acid battery. The overall reaction is shown in Eqn. 4.2 for which it is estimated that the relativistic stabilization of the 6s electrons destabilizes the Pb(IV) species PbO_2 to such an extent that the voltage available from this battery is increased from 0.4V to a much more useful 2.1V.

$$Pb(s) + PbO_2(s) + 2\ H_2SO_4(aq) \rightarrow 2\ ^{\iota}bSO_4(s) + 2\ H_2O(l) \qquad (4.2)$$

The group maximum oxidation state is the maximum for that group given the total number of valence electrons available.

Ga(III) and Ge(IV) compounds respectively are not unusual since the bond energies are large enough to compensate for the increased ionization or promotion energies involved. For the later elements, however, the corresponding energies are perhaps too large to allow for sufficient compensation from the formation of extra bonds.

If we consider the inert pair effect and the matter of group maximum oxidation states of the 4p elements together, we see that, certainly for Groups 15 and 16, there is an alternation in the highest stable oxidation state, for example in Group 15: P(V), As(III), Sb(V), Bi(III). Not surprisingly, this is sometimes referred to as the **alternation effect**.

Comparative trends in oxidation states across the periodic table

Another point to consider in this section is the difference between the trends in oxidation states found in the p-block and those observed in the d- and f-blocks (there are no real trends to consider in the s-block). As we have seen, there is a general trend to lower oxidation states as the groups in the p-block are descended (most notable in Groups 13–16), which can be traced to the inert pair effect, a prominent aspect of which is the weaker bonds formed by the larger elements. In the d- and f-blocks, the reverse tends to be true, at least for the elements on the left-hand side of these blocks. Thus, third-row transition metals and actinides often exhibit higher oxidation states than first-row transition metals and the lanthanides respectively (e.g. Os(VIII) and U(VI) have no counterpart in the chemistry of their lighter group congeners iron and neodymium respectively). This can be traced to the difference in d and f compared to s and p orbitals and relativistic effects.

We can consider the situation in the p-block to be 'normal', explanations for which have been offered above. For the heavier elements of the d and f blocks, however, relativistic effects cause a contraction of the valence (and core) s orbitals, and, to some extent, the p orbitals which leaves the d and f orbitals exposed (this is the indirect relativistic orbital expansion referred to in Section 2.7) with the electrons in these orbitals feeling a reduced effective nuclear charge or smaller effective core potential. They are therefore easier to remove or ionize, and so higher oxidation states are found in the chemistry of these elements. It is also the reason why covalent bond strengths tend to increase down a group in the d-block whereas the reverse is true in the p-block; we will return to this point later. The fact that higher oxidation states are less common as the d- and f-blocks are traversed from left to right simply reflects the increasing effective nuclear charge which means the electrons are, in general, harder to remove.

The trend to lower oxidation states in the p-block is not obvious in the later groups (i.e. Groups 17 and 18) for two reasons. Firstly, for the elements in the top right of the p-block, electrons are too strongly held for high oxidation states to be readily formed, and secondly, the paucity of available chemical data for the radioactive elements At and Rn means that any tendency to form lower oxidation states is unknown.

Comparative stabilities of oxidation states

Much of the discussion so far about trends in oxidation states, such as the inert pair effect and the reluctance of the 4p elements to exhibit the group maximum oxidation state, has been rather qualitative, particularly in terms of the

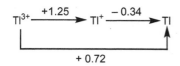

Fig. 4.3 A Latimer diagram for thallium.

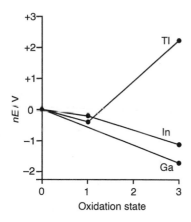

Fig. 4.4 A Frost diagram for gallium, indium, and thallium.

explanations offered to account for these observed trends. More quantitative relationships concerning the stability of oxidation states for a particular element are available, at least under certain conditions, from a consideration of electro-chemical data. This is a big topic and one we can only touch on very briefly here, but it is worth at least a cursory look. Much more detail is given in, for example, Weller, Overton, Rourke, and Armstrong (2018) and other standard texts.

Electrochemical data in aqueous solution for particular chemical species at particular values of pH can be summarized in so-called Latimer, Frost, or Pour-baix diagrams. For systems which contain oxo-cations or the more common oxo-anions (for example, species such as perchlorate, ClO_4^-), Latimer and Frost diagrams vary according to pH (dependence on pH is explicitly part of Pourbaix diagrams) which means that the relative stability of oxidation states also varies with pH, sometimes to a considerable degree. We will not look at that sort of complexity here, and will illustrate the usefulness of Latimer and Frost diagrams only with reference to a very simple example associated with the heavier Group 13 elements.

An example of a Latimer diagram for thallium is shown in Fig. 4.3 which shows the relationship between the expected oxidation states Tl^{3+}, Tl^+, and $Tl(0)$ in aque-ous solution (there is no pH dependence here since we are not dealing with any oxo-species). The numerical values are the standard reduction potentials in volts, i.e. $E^{\ominus}/V$. These data, along with similar data for Ga and In, are shown in the form of a Frost diagram in Fig. 4.4. A Frost diagram plots oxidation state against $nE^{\ominus}/V$ (where n = the number of electrons), and it is important to recognize a number of key points:

1. The reduction potential is related to the free energy according to $\Delta G^{\ominus} = nE^{\ominus}$, so all considerations are thermodynamic and there is no information on any kinetic factors.

2. The most thermodynamically stable oxidation state (for a given pH) is the one that lies lowest on the diagram.

3. The slope of the line joining any two points on the diagram is the standard potential of the couple; the steeper the slope, the higher the potential.

4. An oxidation state is unstable with respect to disproportionation if it lies above a line connecting species on either side.

5. Two species will comproportionate to an intermediate state if the intermediate state lies below a line connecting the two species.

If we look at indium and thallium in particular, it is immediately apparent that Tl(III) is unstable with respect to Tl(I) and, moreover, that Tl(0) and Tl(III) are un-stable with respect to comproportionation to Tl(I). Contrast this with the case for indium where In(III) is the most stable state and In(I) is unstable (albeit only to a small degree) with respect to disproportionation into In(0) and In(III). No data is available for Ga(I), but Ga(III) is clearly more stable than Ga(0). In summary, therefore, the instability of Tl(III) with respect to Tl(I), which was one of the exam-ples given of the inert pair effect, is clearly revealed in these data.

4.3 Element size and coordination numbers

The effect of atomic size is often underestimated or understated when considering the structures of inorganic molecules, but we should bear in mind that larger atoms can and do support larger coordination numbers. We shall see many examples of this as we look at the various classes of compound, but in addition to any structural effects we should remember that, for a given coordination number, larger central atoms are also likely to be more reactive since there is easier access to the element centre.

For ionic structures, the particular structure type adopted can often be understood very much on the basis of the sizes, or relative sizes, of the constituent ions, as we shall see in Chapter 5.

4.4 Bond energies

Homonuclear bonds

Approximate bond energies, which are a measure of bond strength, for homonuclear single bonds for some of the s- and p-block elements are given in Table 4.3, but we must exercise some care in using bond energy data, since values can vary considerably depending on the compound involved (for example, the bond dissociation energy measured for the N–N bond in N_2H_4, N_2F_4 and N_2O_4 is 167, 88, and 57 kJ mol^{-1} respectively). In making detailed comparisons, therefore, it is necessary to compare similar compounds, but we are not concerned here so much with precise energies, but rather with reasonable values from which we can discern any trends which are apparent.

Firstly, it is clear that, in general, bonds become weaker as the groups are descended. This is generally attributed to the fact that the atoms are increasing in size and therefore the orbitals are also increasing in size becoming more diffuse to the point where overlap is poor and bonds are correspondingly

Table 4.3 Homonuclear single bond energies for the s- and p-block elements (kJ mol^{-1})

H–H 432						
Li–Li 105	Be–Be (208)	B–B 293	C–C 346	N–N 167	O–O 142	F–F 155
Na–Na 72	Mg–Mg (129)	Al–Al –	Si–Si 222	P–P 201	S–S 226	Cl–Cl 240
K–K 49	Ca–Ca (105)	Ga–Ga 115	Ge–Ge 188	As–As 146	Se–Se 172	Br–Br 190
Rb–Rb 45	Sr–Sr (84)	In–In 100	Sn–Sn 146	Sb–Sb 121	Te–Te (126)	I–I 149

Values in Table 4.3, and in subsequent Tables containing bond energy data, are taken from Huheey, Keiter, and Keiter (1993), which also provides some useful definitions of bond energies. Values in parentheses are calculated values. No indication of errors is given, which are substantial in some cases.

Note the strength of the H–H bond.

weak. A complementary explanation is that the trend to weaker bonds as groups are descended is a consequence of more weakly bonded valence electrons.

An obvious exception to this generalization occurs for the first-row elements of Groups 15, 16, and 17. Note that whilst B–B and C–C bonds are the strongest of their group, the energies of the N–N, O–O, and F–F single bonds are unusually weak. The simplest explanation for this 'anomaly', as it is often called, has to do with the presence of lone pairs and the small size of the atoms concerned. Thus, each nitrogen, oxygen, and fluorine atom have one, two, and three non-bonded or lone pairs of electrons respectively which are held relatively close to the nucleus (bond pairs are shared between two atoms), and this factor, coupled with the particularly small size of these atoms, leads to appreciable inter-electron repulsions between neighbouring atoms thereby destabilizing the bond. In other words, for sufficient orbital overlap to occur to form a strong element–element bond, the atoms would have to get so close together that substantial repulsions between their non-bonded electron pairs would result. In the cases of boron and carbon, all the electron pairs are bond pairs, so that inter-electron repulsions around the central atom are much less and the atoms are also slightly larger, both factors accounting for the presence of much stronger bonds. For the second and subsequent row elements, the larger size of the atoms means that inter-lone pair repulsions are less important (although see later) with the result that second-row element–element bonds are the strongest in the group for Groups 15–17, with subsequent rows becoming weaker as expected.

Another explanation advanced for the weakness of the N–N, O–O, and F–F single bonds derives from a consideration of the molecular orbitals involved. Thus, if we consider again the weakness of the F–F bond in F_2, this originates, according to the molecular orbital approach, from the presence of filled antibonding F–F π^*-orbitals as shown in Fig. 4.5 which is a simplified molecular orbital energy level diagram for F_2. The reason this is important is that, in general, antibonding orbitals are more destabilizing than bonding orbitals are stabilizing such that the filled F_2 π^*-orbitals do more than just 'cancel out' the filled π-orbitals, i.e. ΔE_a in Fig. 4.5 is larger than ΔE_b. Any corresponding effect for Cl_2, for example, would be much less significant since all the orbitals lie closer together in energy and the difference between ΔE_a and ΔE_b would therefore be less. For a more in-depth discussion of this topic and of molecular orbitals in general, see Winter (2016).

Trends in bond energy across a period are somewhat less marked, but it is clear that, in general, there is a significant increase in bond strength as we move from Group 1 to Group 14, with a slight decrease at Group 15 followed by a roughly constant or modest increase in bond energy as we go from Group 15 to Group 17. For the 2p elements, the drop at Group 15 (i.e. nitrogen) is considerable, and the O–O and F–F bond strengths quoted are slightly lower than that for the N–N bond. The trend from Group 1 to Group 14 results from the increasing

More sophisticated electronic structure calculations point not so much to lone pair–lone pair repulsions but rather to significant Pauli repulsion between σ-orbitals in small first-row elements; Pauli repulsion is addressed later.

For N_2O_4 in particular (where there are no lone pairs present on the nitrogen atoms), another explanation for the weak N–N bond may lie in the fact that the electronegative oxygens lead to δ⁺ charges on each nitrogen and that the close proximity of these δ⁺ charges weakens the bond (see below). A similar explanation can be offered to account for the weakness of Al–Al bonds.

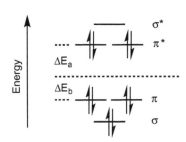

Fig. 4.5 A simplified molecular orbital energy level diagram derived from overlap of the 2p orbitals showing the σ and π molecular orbitals for F_2.

electronegativity of the atoms since the more electropositive atoms to the left of the periodic table, when mutually bonded, will tend to have destabilizing, adjacent δ^+ charges (see marginal notes for N_2O_4 and $R_2Al–AlR_2$). The trend to more electronegative atoms with smaller orbitals and therefore better overlap may also provide an explanation for the observed increasing bond energies. On moving from Group 14 to Group 15, the elements now have lone pairs (at least for the trivalent state), and the reasons discussed above for the weak N–N, O–O, and F–F bonds, where it is most pronounced, are the main reason for the observed decrease in bond energy.

In a compound $R_2Al–AlR_2$, the polarization will be $R(\delta^-)_2Al(\delta^+)–(\delta^+)AlR(\delta^-)_2$. since any R group is likely to be more electronegative than Al.

Heteronuclear bonds

If we now look at heteronuclear bonds, it is apparent that these are generally much stronger than the homonuclear bonds we have just considered, as is evident from the bond energies for the s- and p-block-element fluorides displayed in Table 4.4.

A general explanation for the disparity between heteronuclear and homonuclear bond energies was offered by Pauling who ascribed it to the difference in electronegativity of the elements involved; indeed, the concept of electronegativity largely evolved from a consideration of the differences between the strengths of homonuclear *vs* heteronuclear bonds. Thus, any difference in electronegativity of the elements forming a bond will give rise to a certain degree of ionicity or charge separation (δ^+–δ^-) which will lead to an additional electrostatic attraction (sometimes called an ionic resonance energy) over and above that resulting from the covalent bond itself. Pauling proposed an equation to calculate the bond dissociation energy (*D*, in kJ mol^{-1}) of a heteronuclear bond based on the values of the respective homonuclear bonds and the square of the electronegativity difference between the atoms (Eqn. 4.3) although this argument

Table 4.4 Bond energies for the s- and p-block-element fluorides (kJ mol^{-1})

H–F
565

Li–F	Be–F		B–F	C–F	N–F	O–F	F–F
573	632		613	485	283*	190*	155
Na–F	Mg–F		Al–F	Si–F	P–F	S–F	Cl–F
477	513		583	565	490*	284	142*
K–F	Ca–F		Ga–F	Ge–F	As–F	Se–F	Br–F
490	550		469	452	406	285	187*
Rb–F	Sr–F		In–F	Sn–F	Sb–F	Te–F	I–F
490	553		444	414	402	329	231
Cs–F	Ba–F		Tl–F	Pb–F	Bi–F	Po–F	At–F
502	578		439*	331	297	–	–

The bond energies quoted in Table 4.4 are for the elements in their group maximum oxidation states, except for those indicated with an asterisk. As with Table 4.3, no indication of any variation in values is given which may be considerable (up to ± 50 kJ mol^{-1}) and quite dependent on oxidation state.

is somewhat circular since it is bond dissociation energies which were used to calculate electronegativities in the first place. However, as we saw in Chapter 2, there are many other scales of electronegativity most, if not all, of which are consistent with the Pauling scale.

$$D_{(A-B)} = \tfrac{1}{2}\left[D_{(A-A)} + D_{(B-B)}\right] + 100\left(\chi_A - \chi_B\right)^2 \qquad (4.3)$$

In Eqn. 4.3, Pauling originally gave a proportionality constant of 1, where D was quoted in eV. To convert values of $\chi_A - \chi_B$ in eV to kcal mol^{-1}, a proportionality constant of 23 was later used, but to convert to modern units of kJ mol^{-1}, the constant should be 100.

Whilst Eqn. 4.3 is only approximate, it is a useful guide to the strength of heteronuclear bonds and provides a simple rationale for the high bond energies involving fluorine in particular, although other factors such as π-bonding effects are also important, as we shall see.

More sophisticated treatments of bond energies define a total bond energy, ΔE, according to Eqn. 4.4, where ΔE_{int} is a collection of terms as shown in Eqn. 4.5, and ΔE_{prep} is the energy required to attain a valence state, i.e. a promotion energy.

$$\Delta E = \Delta E_{int} + \Delta E_{prep} \qquad (4.4)$$

$$\Delta E_{int} = \Delta E_{orb} + \Delta E_{elstat} + \Delta E_{Pauli} + \Delta E_{disp} \qquad (4.5)$$

More detail and further discussion relating to Eqns. 4.4 and 4.5 can be found in a 2019 review article by Frenking and Schwerdtfeger and co-workers (*Chem. Rev.*, 2019, **119**, 8781).

In brief, ΔE_{orb} corresponds to orbital overlap, ΔE_{elstat} relates to coulombic attractions (which includes the ionic contribution noted previously), ΔE_{Pauli} is the Pauli repulsion component, and ΔE_{disp} is the (much the weakest of the four) dispersion or van der Waals term. Pauli repulsion is related to the Pauli Exclusion Principle and arises when electron pairs try to occupy the same volume or region of space.

If we look first at the s-block elements in **Group 1** and **Group 2**, it is clear that the bond energies, or, to be more precise, lattice energies since we are dealing here with ionic compounds (see later), are all around 500–600 kJ mol^{-1} which are very considerable values. As we descend Groups 1 and 2, no particular trends are apparent, and the precise values will in any case depend on the details of the ionic crystal structure and are themselves subject to error, but it is clear that the bond energies for the Group 2 fluorides are higher than those for the corresponding Group 1 fluorides which results from the presence of 2+ cations in the case of the former; lattice energies are proportional to the charge of the ions.

Although a lattice energy and a covalent bond energy seem rather different, they are both a measure of the binding energy between atoms so there is no problem in discussing them together.

Large bond energies for the ionic s-block fluorides coupled with the three-dimensional bonding present in ionic solids accounts in large part for the characteristic features of ionic materials such as hardness and high melting points. Note that the lattice energy ΔH_{latt} is proportional to $z^+z^-/r^+ + r^-$, where z is the ionic charge, and r is the ionic radius—something we shall look at in more detail in Chapter 5.

In **Group 13**, the values quoted in Table 4.4 are for compounds in the trivalent state, with the exception of Tl–F, the value for which is derived from thallium monofluoride (bond energies vary with oxidation state, sometimes by a considerable amount, as we shall see), and show a general decrease as we descend the group. This does not mirror the trends in electronegativity whereby we might expect, for example, that the Al–F bond energy would be greater than that for a B–F bond, but we must recognize that size will also be important, and since aluminium is bigger than boron, this will tend to weaken the Al–F bond relative to the B–F bond. Furthermore, since trivalent boron and aluminium compounds often have a vacant orbital, the possibility of B–F and Al–F π-bonding exists with the former expected to be considerably stronger for reasons discussed in the following section. Remember that Eqn. 4.3 estimates how much stronger an A–B bond will be in relation to the mean of A–A and B–B bond energies, but it would be wrong to assume that A–B bond energies are determined solely by element electronegativity differences.

In **Group 14** the general trend is the same as for Group 13 for much the same reasons, and for the heavier elements it would appear that size considerations are probably more important than electronegativity differences such that bond strengths decrease down a group. However, the important difference in this group compared to Group 13 is that the C–F bond is considerably weaker than the Si–F bond (*cf* B–F > Al–F). This is clearly not to be expected on the basis of element size and undoubtedly reflects the greater electronegativity difference between Si and F *vs* C and F in this case. Moreover, with silicon there is also the possibility of Si–F π-bonding due to the availability of low-energy vacant orbitals on silicon, analogues of which are not present for carbon; the nature of these orbitals will be addressed in Chapter 6.

A situation where size and electronegativity difference work together can be seen by considering the carbon-to-halide *vs* the silicon-to-halide bond energies shown in Table 4.5. Thus, there is a progression to weaker bonds as the halogen group is descended since the size of the halogen is increasing (resulting in poorer overlap) and the electronegativity difference is decreasing. Note, however, that in all cases the Si–X (X = halogen) bonds are stronger than the C–X bonds for a given X as discussed previously for the specific case of the fluorides. A similar situation is found for the oxides, where Si–O (452) and P–O (335) single bonds are found to be much stronger than C–O (358) and N–O (201) single bonds respectively.

These comments on π-bonding, i.e. B–F > Al–F but Si–F > C–F, may appear confusing. Where a primary valence orbital is available (as with B and Al), π-bonding is expected to be strongest for the 2p element, but where orbitals of a different type (generally vacant d orbitals or σ^*-orbitals; see Chapter 6) are present, π-bonding is generally greater for the 3p element.

The whole issue of π-bonding in explaining bond strengths (and lengths) to fluorine has been questioned by Gillespie, who has proposed that most element–fluorine bonding should be treated (and can be satisfactorily explained) as ionic in view of the invariably large element-to-fluorine electronegativity difference.

Table 4.5 Heteronuclear carbon and silicon to halogen bond energies (kJ mol^{-1})

C–F	485	C–Cl	327	C–Br	285	C–I	213
Si–F	565	Si–Cl	381	Si–Br	310	Si–I	234

In **Group 15** the situation with the element fluorides is the same as that for Group 14, but we should note that the N–F bond is much weaker than the P–F bond. This is a consequence of both N and F being small elements and having lone pairs, the effect of which was discussed in the case of homonuclear bonds, but the N–F bond is a lot stronger than either the N–N bond (167) or the F–F bond (155) which clearly demonstrates the effect of electronegativity difference. This feature is reflected in the chemistry of the nitrogen halides wherein NF$_3$ is a thermodynamically stable and unreactive molecule in complete contrast to the heavier congener, NCl$_3$, which is extremely unstable and reactive due largely to the small N–Cl bond energy (190 kJ mol^{-1}) which stems from the similar electronegativities of nitrogen and chlorine; P–H, C–I, and S–I bonds tend to be rather weak for similar reasons. Bromides and iodides of nitrogen are even more unstable which also reflects the general trend to weaker bonds as a group is descended.

Weak bonds are not the sole reason for the instability of compounds containing certain bond types. The S–I bond, for example, although fairly weak (~210 kJ mol^{-1}), is certainly respectable, and the instability of compounds containing S–I bonds (binary neutral sulfur-iodides stable at room temperature are not known) is most likely due to the exothermicity of reactions which form elemental S and I.

The value quoted in Table 4.4 for the Cl–F bond (142 kJ mol^{-1}) is actually lower than that for F$_2$ (155 kJ mol^{-1}), but the value for chlorine in this case is derived from the molecule ClF$_5$. A better comparison with F$_2$ is the ClF molecule, for which the Cl–F bond energy is 249 kJ mol^{-1}, i.e. considerably greater.

In **Group 16** and **Group 17** the O–F and F–F bonds are very weak (small atoms, lone pairs, and small or zero electronegativity difference), but there is now a tendency for the element-to-fluorine bond energies to increase as we descend the group. This observation would indicate that it is now probably electronegativity differences which are dominant rather than size differences which is not unexpected, since the elements of these groups are themselves generally smaller than those of the preceding groups such that the size mismatch between them and fluorine is less marked. Alternatively, it may be that with sulfur and chlorine these atoms are sufficiently small for lone pair–lone pair repulsions across the S–F and Cl–F bonds to be important, whereas with tellurium and iodine, for example, such a factor is much less important. The fact that E–F bonds are generally so much stronger that the F–F bond in F$_2$ is one of the reasons why fluorine is so reactive.

As a final point regarding Table 4.4, we should note that, in considering trends across a period, there is, in all cases, a marked decrease in the E–F bond energy as we go from left to right as would be expected from electronegativity considerations.

Another class of heteronuclear bond that we shall consider briefly is that of the s- and p-block elements to hydrogen which are shown in Table 4.6. The general trends we would expect to see are again evident, albeit a little different from those seen for element–fluorine bonds as a result of electronegativity differences. Thus, bond strengths are at their largest at the top right of the table, since this is where the electronegativity difference between the element and hydrogen is at its greatest. Otherwise, bond strengths tend to decrease down a group with a somewhat less marked change across a row with the exception of the elements in the top right, for reasons noted previously. We should note also, of course, that the absence of any non-bonding electron pairs on hydrogen mean that there are no lone pair–lone pair repulsions and no π-bonding effects to consider.

Table 4.6 Bond energies for the s- and p-block element hydrides (kJ mol^{-1})

H–H 432						
Li–H 243	Be–H 226	B–H 381	C–H 411	N–H 386	O–H 458	F–H 565
Na–H 197	Mg–H 211	Al–H 285	Si–H 318	P–H 322	S–H 363	Cl–H 428
K–H 180	Ca–H 159	Ga–H 260	Ge–H 288	As–H 245	Se–H 276	Br–H 362
Rb–H 163	Sr–H –	In–H 243	Sn–H 314	Sb–H 257	Te–H 238	I–H 294
Cs–H 176	Ba–H –	Tl–H 185	Pb–H 180	Bi–H 194	Po–H –	At–H –

Values in Table 4.6 are taken from Emsley (1989), Huheey, Keiter, and Keiter (1993), and for some of the Group 2 and Group 13 bonds, from Downs and Pulham, *Chem. Soc. Rev.*, 1994, 175.

Before we leave a discussion of the strengths of heteronuclear bonds, we should consider the case of the lead(IV) chlorides and alkyls, described in Section 4.2, since there are a number of important general points which emerge.

Recall that for lead chlorides $PbCl_2$ is more stable than $PbCl_4$ (inert pair effect), whereas for the alkyls, $PbEt_4$ is more stable than $PbEt_2$. If we consider first the former observation, part of the reason is due to the relative weakness of the Pb(IV)–Cl bond, and this highlights a general point concerning the bond strengths of lead(II) vs lead(IV) halides, since it is found that Pb–X bonds in lead(II) halides are generally about 50–60 kJ mol⁻¹ stronger than Pb–X bonds in lead(IV) halides. Moreover, the same is true for the respective +2 and +4 oxidation states of tin and germanium and also for indium where the In–X bond strengths for indium(I) halides may be as much as 100 kJ mol⁻¹ greater than those for indium(III) halides (the situation is similar for gallium and thallium).

Clearly, the involvement of the s^2 electron pair in the higher oxidation state of lead is associated with a considerable weakening of the bonds, and this can be traced to the difference in the radial extension of the s and p orbitals of the heavier elements mentioned briefly in Chapter 1. The result of the markedly different sizes of the s and p orbitals for the heavier elements is that s-p hybridization is less efficient, and covalent bonding involving an s-p hybridized state is correspondingly weaker; in the +2 or divalent state, bonding can occur using only p-orbitals. This is, of course, the same argument (slightly restated) as was presented previously in our discussion of the inert pair effect.

The increasingly disparate size of the valence s and p orbitals as the group is descended is illustrated by the values for the Group 14 elements shown in Table 4.7, the origin of which is another consequence of Pauli repulsion. Thus, the special case for 2s vs 2p arises because 2s electrons experience Pauli repulsion from 1s electrons, but there are no 1p electrons to affect (and therefore repel) 2p electrons. For all other rows, valence s and p electrons experience Pauli repulsion from filled inner s and p orbitals, and since p electrons experience a slightly smaller Z^* than s electrons, p orbitals are correspondingly larger in radial extension.

With the above points in mind, we should anticipate that electronegative substituents will result in particularly weak covalent bonding in situations where s-p hybridization is important (i.e. the tetravalent state), since the high partial positive charge induced at the central element centre will tend to increase the

Perhaps surprisingly, Pb(IV)–X bonds, although weaker than Pb(II)–X bonds, are significantly shorter, which is due to the smaller size of the Pb(IV) centre.

Table 4.7 Orbital sizes (r_{max} in pm) of the ns and np orbitals for the Group 14 elements

	r_{max} ns (pm)	r_{max} np (pm)
C	65	64
Si	95	115
Ge	95	119
Sn	110	137
Pb	107	140

More detail can be found in W. Kutzelnigg (Angew.Chem., 1984, **23**, 272).

There are other reasons for the instability of the heavier halides as mentioned previously—notably the increasing size of the halide and the progressive ease of oxidation by a high oxidation state element centre.

difference in radial extension between the s and p orbitals making s-p hybridization even less efficient. This is undoubtedly an important part of the reason for the instability of the halides of lead(IV) described above, and should be most marked in the case of the fluoride, although we should expect that any weakening of the covalent part is likely to be offset by a strong ionic contribution to the bond energy which would account for the stability of PbF_4 vs the heavier Pb(IV) halides. This is, in fact, a general observation for the heavier p-block elements; high oxidation states involving removal of the s^2 pair are observed for fluorides but are much less common for the heavier halides.

In considering the case of the lead alkyls, the much lower electronegativity of the carbon in the alkyl groups will not result in differentially much weaker bonds for the +4 oxidation state, and there is likely to be much less of a difference in the Pb–C bond energies for the lead(II) and lead(IV) alkyls (in fact, calculations indicate that Pb–C bonds are slightly stronger in Pb(IV) compounds compared to those of Pb(II)). As a consequence, lead will tend to form four bonds rather than two, which is also evident from the calculated exothermicity of the disproportionation reaction shown in Eqn. 4.6; the corresponding situation for the lead fluorides is the opposite, i.e. the disproportionation of PbF_2 is endothermic.

Disproportionation reactions like that shown in Eqn. 4.6 are similarly exothermic for many Group 13 and 14 alkyls and hydrides. Moreover, as we noted briefly above, high formal oxidation state alkyl and aryl compounds are also seen in Groups 15 ($BiMe_5$) and 16 ($TeMe_6$).

$$2\,Pb(II)Et_2 \rightarrow Pb(0) + Pb(IV)Et_4 \qquad (4.6)$$

As we conclude this subsection, it is clear that there is much to consider in attempting to understand and rationalize trends in bond energies, with many factors at work, sometimes acting in concert and sometimes not.

Multiple bonds

Finally, we should consider the trends in the strengths of multiple bonds. Multiple bonding is common amongst the first-row (2p) elements and is a very important aspect of their chemistry. It is much less common amongst the second and subsequent row elements, however, and we can begin to appreciate the reason for this from a consideration of the various bond energies associated with N_2 and P_2 fragments shown in Table 4.8.

Table 4.8 Single and multiple bond energies for dinitrogen and diphosphorus units (kJ mol^{-1})

N–N	167	P–P	201
N=N	418	P=P	310
N≡N	942	P≡P	481

Several points are important. Firstly, note that the N≡N triple bond is extremely strong (only the C≡O triple bond in carbon monoxide is stronger) and almost twice as strong as the P≡P triple bond. Secondly, although the N=N double bond is stronger than the P=P double bond, the more important point is that the N=N double bond is considerably more than twice the strength of the N–N single bond whereas the P=P double bond is much less than twice the strength of the P–P single bond. This second observation can be seen as a result of the particular weakness of the N-N single bond, for reasons already discussed, with the important consequence that in the chemistry of these elements, phosphorus much prefers to form two single bonds rather than one double bond, whilst the opposite is true for nitrogen.

In seeking an explanation for the varying stability of $p\pi$-$p\pi$ multiple bonds between the p-block elements, we must look further at the strengths of the various bonds involved, and an illuminating way in which this can be achieved (specifically for double bonds in this case) is to examine what are called σ and π bond increments for particular pairs of atoms. Table 4.9 lists such increments for several pairs of elements (nitrogen and phosphorus included), and the values are defined such that the first value for each particular bond is that for a typical element–element single or σ-bond; for example, the value of 335 kJ mol^{-1} for carbon is that of the C-C bond in ethane. The second value is the difference in energy between a double or $\sigma + \pi$-bond and that of the single bond given by the first value, and this can be taken as a measure of the strength of the π-bond. For carbon, the difference is that between the bond energy of the C=C double bond in ethene (630 kJ mol^{-1}) and the single C-C bond in ethane.

Clearly the σ-bond in ethane is not the same as the σ-bond in ethene, but the differences are likely to be small; moreover, it is broad comparisons between different types of bonds that are important, rather than a means of quantitatively predicting particular bond energies.

Table 4.9 Approximate bond increments (kJ mol^{-1}) for σ/π-bonds for various element–element pairs

C–C	N–N	O–O		C–O	N–O
335/295	160/395	145/350		335/380	190/370
Si–Si	P–P	S–S		Si–O	P–O
195/120	200/145	270/155		420/170	335/150
Ge–Ge	As–As	Se–Se		C–S	S–O
165/110	175/120	210/125		280/265	275/250

Values in Table 4.9 are taken from W. Kutzelnigg (*Angew.Chem.*, 1984, **23**, 272). There are slight differences between some of the values in Tables 4.9, 4.8, and 4.3 which reflect the use of different sources but, as was mentioned earlier, bond energies depend to some extent on the particular molecule.

If we first consider homonuclear bonds, an important point to emerge from Table 4.9 is that the π-bond increment for the first row or 2p elements is considerably greater than that for the elements of the second and subsequent rows. In other words, π-bond strengths are much larger for the first-row elements and by a factor of between two and three. The standard explanation for this feature is that the second and subsequent row elements, being larger, have correspondingly larger and therefore more diffuse orbitals. Overlap between these bigger orbitals is poorer which results in weaker bonds, particularly in the case of π-bonds where the atomic orbital overlap is side-on rather than the head-on overlap of σ orbitals. When considering the disparity in the prevalence of π-bonding between the 2p elements and those of subsequent rows however, another aspect which we must consider is the weakness of the N-N and O-O single bonds (like the F-F bond). As a result, P-P and S-S single bonds are considerably stronger than N-N and O-O bonds respectively. In considering both of these factors, it is therefore apparent that for Groups 15 and 16, as the values in Table 4.9 clearly indicate, the combination of a $\sigma + \pi$-bond is more stable than

More sophisticated computational studies indicate that reduced $p\pi$-$p\pi$ overlap for the heavier p-block elements is an oversimplification. π-bond energies certainly decrease for the heavier elements, but so do σ-bond energies. The special stability of $2p\pi$-$2p\pi$-bonding in elements and compounds is traced both to reduced Pauli repulsion for 2p electrons (in the absence of any 1p orbitals) coupled with a weakening of the $2p\sigma$-$2p\sigma$-bonding which is due to Pauli repulsion.

two σ-bonds for the first-row elements, whereas the opposite is true for the elements of the second and subsequent rows.

In Group 14 the comparison is a little different in that the strength of the C–C single bond is greater than that of the Si–Si bond since there are no non-bonded electron pairs associated with the carbon atoms. The absence of an extensive chemistry of multiply bonded silicon compounds is due, as before, to the weakness of π-bonding for this element (two σ-bonds are stronger than one σ and one π-bond), and it is noteworthy that Si–Si π-bonding is the weakest of the 3p elements consistent with the larger size of the silicon atom (*vs* P or S). The preponderance of π-bonding in carbon chemistry, however, is not associated with any weakness of the C–C σ-bond, but we may note that much of the extensive and rather special organic chemistry of carbon is the result of the similarity in the strengths of the C–C σ-bonds and π-bonds. With this final point in mind, it is also clear from the values for Group 14 in Table 4.9 that, in the absence of factors which weaken the first row σ-bonds, the ratio of the σ/π-bond strengths drops markedly between the first and second row (0.88 *vs* 0.61) but then remains nearly constant for the second and third row (0.61 *vs* 0.66).

An examination of σ and π increments for heteronuclear bonds is also instructive. For example, Table 4.9 reveals that the π-bond increment is large where both elements are from the first row, although the σ-bond strengths are also large where one of the elements is carbon. Heteronuclear bonds where a second-row element is involved, for example Si–O and P–O, are similar to the homonuclear second-row element cases in having a small π increment, but we should note the particular case of sulfur wherein the σ and π increments for both C–S and S–O bonds are about equal. This is mainly the result of a large π increment (as opposed to weak σ-bonding, *cf* C–C), and in this sense, sulfur is acting more like a first-row element as far as π-bonding is concerned. This may be traced to the relatively small size of the sulfur atom and the consequent better π overlap, and it is therefore not surprising that pπ–pπ multiple bonding is important in the chemistry of this element. Thus, we might expect that the diatomic molecule S_2 would be a stable form of sulfur, and indeed, the strength of the S=S double bond in S_2 (425 kJ mol⁻¹) is only slightly less than that of the O=O double bond in O_2 (495 kJ mol⁻¹). Note, however, that the large π increment for oxygen favours a doubly bonded structure over a structure with single bonds, whereas the reverse is true in the case of sulfur. This latter point is reflected in the exothermicity of the reaction shown in Eqn. 4.7.

$$4 \, S_2 \rightarrow S_8 \quad \Delta H = -520 \text{ kJ mol}^{-1} \tag{4.7}$$

We are now in a position to understand one of the main points raised in Chapter 3 when we looked at the structures of the elements which is that allotropes containing element–element multiple bonding (graphite, C_{60}, N_2, O_2, etc.) have no counterparts in the known allotropes of their heavier congeners; element–element single bonding is much preferred. We should point out, however, that our arguments are thermodynamic. Kinetic factors can and do allow for the isolation and characterization of thermodynamically less stable or metastable allotropes (under ambient conditions, graphite is more stable than diamond, for example) but it is probably fair to say that allotropes of the heavier elements involving multiple bonding are not even kinetically stable under normal conditions.

Bonding between p-block elements *vs* d- and f-block elements

Before we leave this section it is worth noting a further difference between p-block chemistry and that of the d-block elements. As we have seen, bond energies generally decrease as groups in the p-block are descended, but the opposite is often true for the d-block elements with bonds to third-row elements tending to be the strongest. An explanation for this derives from that given in Section 4.2 concerning trends in oxidation number. Thus, the relativistic stabilization and contraction of the s and p orbitals for the third row or 5d elements leaves the d orbitals more exposed and so overlap is better and bonds are stronger. A similar situation is found for the f-block elements where covalent bonding is much more prevalent in the chemistry of the actinides. This is the indirect relativistic orbital expansion point again.

A useful illustration can be found in the case of the tetrafluorides of titanium, zirconium, and hafnium. Recall that in Group 14 the promotion energies are increasing down the group and the bond strengths are decreasing which leads to the appearance of an inert pair effect. For the Group 4 elements, the promotion energies (s^2d^2 to s^1d^3 in this case) are also largest for the heaviest element (Ti, 79; Zr, 58; Hf, 169 kJ mol^{-1}), but the M–F bond energies increase down the group (Ti–F, 588; Zr–F, 647; Hf–F, 676 kJ mol^{-1}) so we neither observe nor should we expect an analogue of the inert pair effect in Group 4.

4.5 The van Arkel–Ketelaar triangle

Before looking at the compounds of the s- and p-block elements in detail (which we will do in Chapter 5), it will be useful to consider the general types of compound we will expect, and this, as we shall see, can often be rationalized on the basis of electronegativities, χ, and electronegativity differences, $\Delta\chi$. Thus, compounds with a large $\Delta\chi$ will tend to be ionic, and those with medium $\Delta\chi$ are likely to form polar covalent bonds and therefore have polymeric or macromolecular structures. For compounds with a small $\Delta\chi$, the type of compound will depend on whether the elements have a low or high electronegativity. In the former case, the compounds are likely to be metallic for much the same reason that elements with low electronegativity are metals. In the latter case we would expect relatively non-polar molecular species.

A useful way to visualize these ideas is with the so-called ***van Arkel–Ketelaar triangle*** or element triangle, shown in Fig. 4.6, in which the vertices are labelled metallic, covalent, and ionic in such a way that the vertical and horizontal axes represent electronegativity difference, $\Delta\chi$, and absolute electronegativity, $\Sigma\chi$, respectively, and are defined such that, for a molecule AB where B is the more electronegative, $\Delta\chi = \chi_B - \chi_A$ and $\Sigma\chi = (\chi_A + \chi_B)/2$.

Fig. 4.6 is derived from the work of Allen (*J. Am. Chem. Soc.*, 1992, **114**, 1510) and Jensen (*J. Chem. Educ.*, 1995, **72**, 395), building on the original work of van Arkel and Ketelaar published in the 1940s. Allen electronegativities have been used for the elements and compounds displayed. Other, more recent approaches (see Keeler and Wothers (2008)) use orbital energies rather than electronegativities, but the essential features of the diagram remain unchanged.

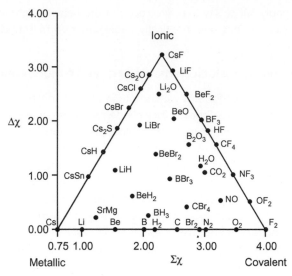

Fig. 4.6 A van Arkel–Ketalaar triangle showing the position of selected compounds and elements.

All discussion is for compounds at atmospheric pressure; metallicity increases with pressure, as noted previously for diiodine and phosphorus. We should also reiterate that electrical conductivity is not only a property of metals; some metal oxides and other solid-state materials have partially filled bands and exhibit metallic conductivity.

The metallic-covalent edge of the van Arkel–Ketelaar triangle comprises the elements themselves, since $\Delta\chi$ is zero, and ranges from metals on the left to covalent species such as F_2 on the right for reasons addressed in Section 2.5 in Chapter 2. An important subcategory in this section is the metalloids such as boron and graphitic carbon, but also including silicon, germanium, arsenic, and antimony. In all cases, covalent bonding resulting from overlap of orbitals is a characteristic feature, and it has been suggested by Allen and Burdett that the specific term 'metallic bond' actually be dropped from inorganic texts since whilst there is certainly a metallic state with defining properties such as electrical conductivity, the bonding in metals is covalent in nature and does not differ in any fundamental way from other types of covalent bonding in which electrons are shared. We shall return to this again in Chapter 8.

The ionic-covalent edge is characterized by decreasing $\Delta\chi$ which is nicely illustrated with the fluorides of the elements. Thus, CsF and LiF are ionic with large $\Delta\chi$ whereas NF_3 and CF_4 both have small $\Delta\chi$ and are covalent; compounds with intermediate $\Delta\chi$ such as BeF_2 are polymeric or macromolecular solids. A similar trend is seen with the element oxides and both classes are described in detail in the next chapter.

The metallic-ionic edge is characterized by increasing $\Delta\chi$, and is illustrated in Fig. 4.6 with compounds of caesium ranging from caesium itself through the metallic CsSn to the ionic compounds Cs_2O and CsF.

The examples examined so far were chosen to lie along the edges of the triangle in order to illustrate the extremes of metallic, covalent, and ionic bonding, but many (in fact, most) compounds will have values of $\Delta\chi$ and $\Sigma\chi$ which place them within the area of the triangle. This is illustrated, albeit only for a limited selection of compounds, in Fig. 4.6 where any compound can be plotted as a function of $\Delta\chi$ and $\Sigma\chi$. Notice, for example, that as for the fluorides which lie along the ionic to covalent edge, the oxides and bromides lie along straight lines which are parallel to this edge but are within the triangle since $\Delta\chi$ and $\Sigma\chi$ values are less in the latter two cases. Similarly, just as caesium compounds lie along the ionic to metallic edge, compounds of less electropositive lithium, beryllium, and boron lie within the triangle on lines parallel to this ionic-metallic edge. Note that because of the way the diagram has been constructed, compounds AB are written with B more electronegative than A so that for any given class of compound where B remains the same and A is varied, such compounds will lie along a line parallel to the ionic-covalent edge. Likewise, for AB where we look at a range of B for a given A, the compounds will lie on a line parallel to the ionic-metallic edge.

A much more detailed account of bonding in terms of the van Arkel–Ketelaar or element triangle can be found in Alcock (1990).

In fact, if we look at a very large number of compounds we can construct a more general plot which clearly reveals the areas of ionic, covalent, and metallic bonding, and this is shown in Fig. 4.7. Furthermore, more detailed quantitative studies have shown that if the nature of the bonding in particular compounds is established according to other criteria, this provides a means of evaluating different scales of electronegativity.

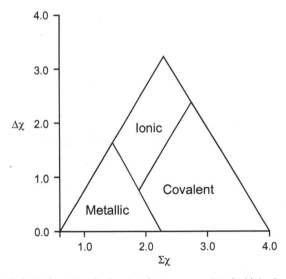

Fig. 4.7 A van Arkel–Ketalaar triangle showing the areas associated with ionic, covalent, and metallic bonding.

As we saw in Chapter 2 (see Fig. 2.5), the metalloid region for the elements is defined by a quite narrow region of electronegativity. With regard to compounds, a particularly important class are the binaries of elements in or near the metalloid region, such as GaAs, InSb, and SiC, together with the binaries for which the element electronegativities are themselves outside the metalloid range, but where the average value, $\Sigma\chi$, lies within this range and where $\Delta\chi$ is sufficiently small for covalent rather than ionic bonding to be present; examples are AlP, GaP, and InP. We can therefore define a metalloid region in the element triangle which is shown in Fig. 4.8.

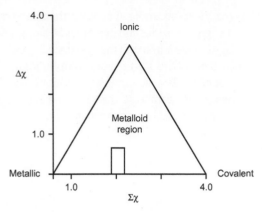

Fig. 4.8 A simplified van Arkel–Ketelaar triangle showing the metalloid region defined by values of $\Delta\chi$ and $\Sigma\chi$.

The treatment outlined above can be taken a step further if we recognize that in addition to metallic, covalent, and ionic bonding, we should also consider van der Waals interactions between discrete covalent molecules in the solid state. Such consideration leads us to the so-called element tetrahedron shown in Fig. 4.9 in which each of the four vertices represents the four bonding types. We can take a compound like solid CI_4 as an example of a material close to the van der Waals vertex wherein the carbon and iodine atoms are linked by covalent C–I bonds whereas the CI_4 molecules are held together by much weaker van der Waals interactions. We will not look at this element tetrahedron in any great detail (the ionic-covalent-metallic face is the same as the van Arkel–Ketelaar triangle, of course), but it is informative to comment on compounds which lie along the three edges which join at the van der Waals vertex.

The element tetrahedron with further informative examples is discussed in more detail by Laing (*Education in Chemistry*, 1993, 160].

A material mid-way along the van der Waals–covalent edge might be selenium, wherein the intermolecular Se···Se interactions between the helical Se_n chains are sufficiently short to suggest the onset of some degree of covalency in addition to any van der Waals bonding (Se–Se = 2.37 Å, Se···Se = 3.44 Å); solid I_2 would be another example, and these types of close interaction which have a degree of covalency are termed ***secondary bonding***, to which we will return in Chapter 6. Gallium is a good example of a material part way along the van der Waals–metallic edge since gallium has definite metallic properties, but the presence of discrete Ga_2 units (see Section 3.3) is suggestive of diatomic units linked by weaker bonding. Finally, as an example along the van der Waals–ionic edge, we might consider $AlBr_3$ which is a low-melting solid of close-packed bromines with aluminium atoms in one sixth of the tetrahedral holes, and which readily sublimes to give Al_2Br_6 molecules (more on close-packed spheres and tetrahedral holes in Chapter 5).

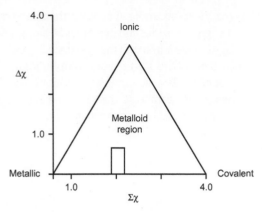

Fig. 4.9 The element tetrahedron.

We might also consider where to place compounds for which intermolecular hydrogen bonding is important such as H_2O (hydrogen bonding is considered again in Chapter 5). Hydrogen bonding is not so much a type of bonding of its own, it is more something which embodies components of van der Waals, ionic, and covalent bonding; somewhere in the element tetrahedron close to the ionic–covalent–van der Waals face would therefore seem appropriate.

Having explored some general aspects, it is now time to look at s- and p-block element compounds and their properties in more detail.

Exercises

1. What is the oxidation state and valence for the first written element in the following compounds: B_2F_4, $[XeF_5]^-$, SO_2Cl_2, $[NO_2]^+$?

Account for the following observations:

1. Disulfur, S_2, is not an allotrope of elemental sulfur under ambient conditions.

2. The perbromate ion, $[BrO_4]^-$, is a stronger oxidizing agent than either perchlorate, $[ClO_4]^-$, or periodate, $[IO_4]^-$.

3. ClF_5 is a strong oxidizing agent.

4. Thallium tri-iodide, TlI_3, exists as the thallium(I) salt $Tl^+[I_3]^-$ rather than as a salt containing Tl(III) and three iodide anions.

5. PCl_5 and $SbCl_5$ show no tendency to lose Cl_2 and form PCl_3 and $SbCl_3$ at room temperature, whereas $AsCl_5$ decomposes to $AsCl_3$ and Cl_2 above –50°C.

6. SnO_2 is easily prepared by oxidation of elemental tin, whereas the formation of PbO_2 requires much stronger oxidizing agents.

Compounds of the s- and p-block elements

5.1 Element halides

The halides are the first class of compound we will look at in detail, and it will soon be clear that it is a large and diverse group, particularly in terms of the variety of structural types encountered. Many of the ideas which we develop here will also be of use when looking at other classes of compound, and we shall start with a broad overview of the basic structural types.

Fluorides

We will look first at the fluorides of the s- and p-block elements in their highest, or group, oxidation states. These are shown in Table 5.1 together with a classification based on the type of structure they adopt under ambient conditions.

We note first that the formulae are of the form EF_n, where $n = 1$–7 according to the group, and that there is a definite progression from ionic materials in the bottom left of the table to molecular covalent species in the top right, with polymeric or macromolecular solids forming an approximately diagonal band in between. We can account for this general arrangement in a straightforward manner by considering the electronegativity differences, $\Delta\chi$, between the elements, and

The known fluorides of nitrogen, oxygen, chlorine, bromine, krypton, and xenon, i.e. NF_3, OF_2, Cl/BrF_n ($n = 1, 3, 5$), KrF_2, and XeF_n ($n = 2, 4, 6$), are not shown in Table 5.1 since the elements in these compounds are not in their highest or group oxidation state, there being no examples for these elements which are.

In Chapters 2 and 4 we noted that to an extent the electronegativity of an element is dependent on its oxidation state. We shall meet this again later in relation to specific examples of compounds, but the differences are not such as to detract from the general point reflected in Table 5.1 and the associated discussion.

Table 5.1 A structural classification of the s- and p-block element fluorides in their group oxidation state. The block containing those compounds classified as polymeric is shaded for clarity

1	2	13	14	15	16	17
LiF	BeF_2	BF_3	CF_4			
NaF	MgF_2	AlF_3	SiF_4	PF_5	SF_6	
KF	CaF_2	GaF_3	GeF_4	AsF_5	SeF_6	
RbF	SrF_2	InF_3	SnF_4	SbF_5	TeF_6	IF_7
CsF	BaF_2	TlF_3	PbF_4	BiF_5		

ionic   polymeric   molecular
                                            covalent

we shall look at each in turn. We have already considered the consequences of $\Delta\chi$ in the van Arkel–Ketalaar triangle in Section 4.5.

Following on from the diagonal band mentioned above, this is probably a good place to introduce what is known as the ***diagonal relationship***. This relationship (that is, from upper left to lower right) is often invoked to account for similarities in the chemistry and properties of pairs of elements that are diagonally related in the periodic table; common examples are Li and Mg, B and Si, C and P. Clearly the stoichiometries of compounds are different for respective pairs since they are in different groups and therefore have different numbers of valence electrons, but the point is that these diagonally related pairs of elements have similar electronegativities and therefore similar electronegativity differences with other elements with which they form compounds. An example would be the similarity between organolithium reagents (RLi) and Grignard reagents (RMgX) and their uses in organic synthesis. There is also a book called *Phosphorus: The Carbon Copy* (1998) which highlights the similarities between phosphorus and carbon chemistry.

Returning to element fluorides, fluorine is the most electronegative element in the periodic table (with the exception of some of the noble gases which are rather special cases). Thus, in forming compounds with the electropositive elements on the left of the periodic table, particularly those near the bottom left, there will be a considerable degree of charge transfer resulting in the formation of the negatively charged fluoride anion, F^-, and a positively charged element cation, E^+ (more generally E^{n+} where n is the number of valence electrons). The resulting materials formed are therefore ionic and adopt solid, three-dimensional structures, or lattices, in which cations are surrounded by anions and vice versa. The bonding is predominantly electrostatic in nature, the lattice being held together largely as a result of the opposite charges on the ions. In general, this type of bonding is strong and ionic solids are characterized, as we noted in Chapter 4, by having high melting and boiling points and by dissolving only in polar solvents such as water. Typical ionic structures for **Group 1** and **Group 2** fluorides are those of sodium fluoride and calcium difluoride shown in Fig. 5.1. We shall say more about ionic solids in general in Section 5.2.

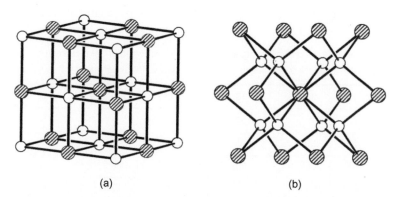

(a)   (b)

Fig. 5.1 Representations of the structures of (a) sodium fluoride, NaF, and (b) calcium difluoride, CaF_2. In both cases, the fluoride anions are the unfilled circles.

If we look now at the molecular covalent species in the top right of Table 5.1, it is clear that all are compounds formed by elements with a similar electronegativity to fluorine. The result is that there is only a small degree of charge transfer on bond formation and the element–fluorine bonds are relatively non-polar, covalent bonds. Thus, there is little or no aggregation between molecules and the compounds are typically gases, low-boiling liquids, or low-melting solids held together by van der Waals interactions and which are soluble in non-polar solvents. Some examples are shown Fig. 5.2.

The intermediate class of compound, labelled polymeric in Table 5.1, consists of those in which the atoms have a medium difference in electronegativity. The difference is not sufficient for the formation of ions and ionic bonding, but the covalent bonds formed are quite polar with large partial positive and negative charges, i.e. $\delta^+E–F\delta^-$. The effect of this in the solid state is that bridging interactions tend to occur between atoms of opposite partial charge leading to polymeric or macromolecular structures. Whilst aggregation or polymerization is a consequence of charge separation or polar bonds, we should not assume that the resulting intermolecular interactions between monomeric units are entirely electrostatic. All of the bonding is predominantly covalent, and we can think of it as arising from the polymerization of monomers, a requirement for which is the presence of low-lying vacant orbitals on the central atom. These may be vacant primary valence orbitals, as in Group 13 chemistry, or other types of orbital the nature of which we will address in Chapter 7. An example of this class is the structure of SbF_5 shown in Fig. 5.3, which exists in a linear polymeric form in the liquid state and as tetramers in the solid.

It should not be assumed that the above three classes of compound are completely distinct from each other, since we cannot categorize bonds as simply ionic, polar covalent, or covalent; all bonds lie on a continuum from highly ionic to purely covalent, as noted in Section 4.5. There is certainly some overlap and occasional resulting ambiguity, but the categories are useful, nevertheless.

So far, we have seen that the broad classification of the element fluorides, at least for elements in their highest oxidation states, can largely be rationalized on the basis of electronegativity differences, $\Delta\chi$. We shall now look at some of the lower oxidation state fluorides and see how these ideas can be developed to understand their structures as well.

Remember that although the electronegativity differences, $\Delta\chi$, between the elements is small, the absolute values, $\Sigma\chi$, are large, and so we are dealing with covalent molecules rather than metallic species.

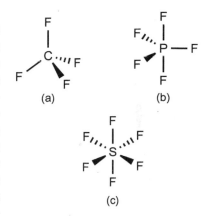

Fig. 5.2 Diagrams of the structures of (a) CF_4, (b) PF_5, and (c) SF_6.

Electronegativity differences are not the *only* factor to consider. For example, aluminium trifluoride, AlF_3, is polymeric, whereas silicon tetrafluoride, SiF_4, is a monomer, despite the electronegativity differences with fluorine being similar. AlF_3 is polymeric due to the presence of a vacant 3p orbital on the aluminium centre in monomeric AlF_3 which results in the facile formation of extra bonds. There is no such vacant orbital in SiF_4.

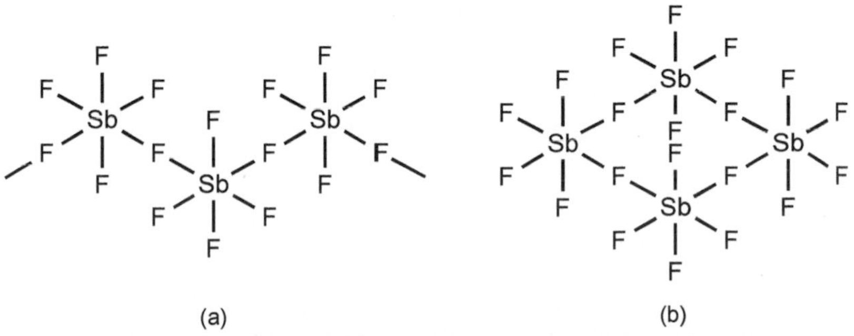

Fig. 5.3 Representations of the structure of SbF_5 in (a) the liquid and (b) the solid state.

There are no stable lower oxidation state fluorides in Group 2 (remember the point made earlier in Section 4.2), but in **Group 13** we encounter examples of boron(II) and thallium(I), B_2F_4 and TlF respectively. The former is molecular covalent (with a B–B bond) and the latter ionic in accord with their electronegativity differences with fluorine. Furthermore, it is interesting to compare the thallium fluorides TlF_3 and TlF. We should expect the former to be 'more covalent' than the latter, i.e. TlF should be the most ionic, since Tl(III) is more electronegative than the Tl(I) (see Section 2.5 in Chapter 2 and recall that the Pauling electronegativities for Tl(III) and Tl(I) are quite different, Tl(III) being the larger). Whilst this resulting difference in $\Delta\chi$ for Tl(III)F_3 and Tl(I)F is not obviously reflected in the structures and physical properties of the solids, it is noteworthy that TlF_3 hydrolyses readily in water to give $Tl(OH)_3$ and HF, whereas TlF dissolves to give aqua ions, the former being more a feature of covalent materials, and the latter a feature of ionic compounds.

In **Group 14** the other common oxidation state is +2, at least for the heavier elements, and the element difluorides and tetrafluorides are interesting classes of compounds which offer a useful illustration of the concepts we have developed so far to rationalize structure. Recall (Table 5.1) that GeF_4 is molecular covalent (Fig 5.4(a)), whereas SnF_4 and PbF_4 are polymeric. We can partly rationalize this observation on the basis of electronegativity differences, but we should also note that the larger tin and lead atoms are also better able to support larger coordination numbers (four for Ge in GeF_4 and six for Sn and Pb in the solid-state structures of SnF_4 and PbF_4).

GeF_2, in contrast to GeF_4, is polymeric as shown in Fig. 5.4(b), and we can also use the idea of electronegativity differences to account for this observation (In part) if we recognize that Ge(II) is likely to be less electronegative than Ge(IV). We commented on a similar situation for Tl(III) vs Tl(I), but the consequences are more dramatic here since the polarity of the Ge–F bond in GeF_2 is now sufficiently large to result in a polymeric structure, whereas in GeF_4 the polarity is small enough for a covalent molecular structure to be favoured.

We should therefore expect that for the fluorides of a given element (and, as we shall see, element halides and oxides in general) there will be a trend from covalent to polymeric to ionic as the central element oxidation state, and hence electronegativity, decreases. This is illustrated not so much with SnF_2 which is also polymeric, but with PbF_2 which adopts typically ionic structures; the α form

Rather than using the argument that Ge(IV) is more electronegative than Ge(II), we can also offer an explanation in terms of Ge(IV) being more polarizing than Ge(II). Smaller, more highly charged centres are better at attracting (polarizing) electron density towards themselves, resulting in more covalent bonds. **Polarization** and **polarizability** are considered again in Chapter 6 when we look at hard and soft acids and bases.

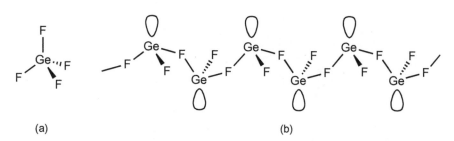

(a)   (b)

Fig. 5.4 The structure of (a) GeF_4 and (b) part of the polymeric structure of GeF_2, including a representation of the lone pair on each Ge.

is analogous to $BaCl_2$ and the β form has the fluorite structure, the coordination numbers around the lead atom being nine and eight respectively.

Nevertheless, whilst electronegativity differences *are* a significant factor, we must not overstate their importance, or ignore other factors such as element size and the nature of any vacant orbitals which are present (and in some cases, the extent of any π-bonding). Thus, GeF_2 is also more likely to be polymeric than GeF_4, since the former is only two-coordinate (as a monomer, and clearly coordinatively unsaturated) whereas the latter is four-coordinate. This factor coupled with the larger size of Ge(II) *vs* Ge(IV) would enable Ge(II) to expand its coordination number, i.e. polymerize, more readily than Ge(IV). Furthermore, the presence of a vacant p orbital in monomeric GeF_2 would make this species particularly Lewis acidic and therefore prone to polymerization; the analogy between AlF_3 and SiF_4 mentioned earlier is relevant here, and we will return to this matter again in Chapter 6. Another factor to note is that whilst monomeric GeF_4 is tetrahedral and has no dipole moment, monomeric GeF_2, for which a bent structure is expected, would have a dipole moment which would also facilitate aggregation.

In **Group 15** all elements have a stable trifluoride for which we see the expected progression in structural types. Thus NF_3, PF_3, and AsF_3 are all molecular, whereas SbF_3 is polymeric. BiF_3 is best described as ionic and contrasts with the polymeric BiF_5, this observation again being consistent with the lower electronegativity or polarizability of Bi(III) *vs* Bi(V).

In **Group 16** there are three common oxidation states, and in addition to SF_6, the fluorides SF_2, S_2F_2 (two isomers), S_2F_4, SF_4, and S_2F_{10} are also known, all of which are molecular. The chemistry of selenium is more limited and, besides SeF_6, only SeF_4 is stable. Selenium tetrafluoride is molecular and has the same structure as SF_4 (Fig. 5.5(a)), but the analogous tellurium compound, TeF_4, is polymeric with a structure shown in Fig. 5.5(b).

Note that coordination numbers are generally lower for lower oxidation states. In polymeric GeF_2, the germanium atom is only three-coordinate whilst in monomeric GeF_4 it is four-coordinate. This is a result of the presence of a lone pair of electrons in the former which occupies one of the valence orbitals leaving only three for bonding (formally two normal Ge–F bonds and one dative F→Ge bond in this case).

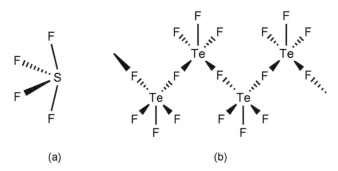

Fig. 5.5 The structure of (a) SF_4 and (b) part of the polymeric structure of TeF_4.

In Table 5.1, the only fluoride of **Group 17** is IF_7, since this is the only known neutral interhalogen compound with a central element in the formal +7 oxidation state; there are no **Group 18** halides in the +8 oxidation state. We should note, however, that a number of neutral lower oxidation state Group 17 and 18 fluorides are known. Examples are EF_3 and EF_5 (E = Cl, Br, I) together with KrF_2 and XeF_n (n = 2, 4, 6), all of which are molecular; the structures of ClF_3 and IF_5 are shown in Fig. 5.6.

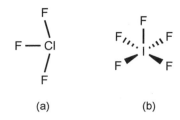

Fig. 5.6 The structure of (a) ClF_3 and (b) IF_5.

In line with the work of Gillespie discussed in previous chapters in terms of an ionic model for most element fluorides, to talk about top right element fluorides as containing weakly polar E–F bonds may be rather inaccurate. The molecular nature of these compounds would then be traced to the fact that the small element centres are essentially coordinatively saturated which prevents polymerization.

Before moving on to look at other halide compounds, we should again stress the tendency of elements to form polymeric and ionic fluorides as groups are descended and as oxidation states are reduced, both of which result from increasing electronegativity differences and, in the case of descending groups, from the fact that element sizes are increasing.

Chlorides

Let us now consider compounds formed by the other halides. We should expect, since chlorine, bromine, and iodine are less electronegative than fluorine, that there will be fewer ionic compounds formed and that there will also be a trend towards compounds with intermetallic or metallic alloy properties, as we shall see with some of the iodides.

Table 5.2 lists the common chlorides of the s- and p-block elements, although in contrast to Table 5.1, some lower oxidation state compounds are included as well, since the number of known chlorides of the elements in their group oxidation state is not large, particularly for the right-hand side elements. This trend is even more marked for the bromides and iodides, and is a result of the fact that the heavier halides become progressively easier to oxidize and are therefore correspondingly less able to stabilize high formal oxidation states.

The chlorides of **Group 1** and **Group 2** are ionic as expected with the exception of $BeCl_2$ which has a chloride-bridged, linear polymeric structure as shown in Fig. 5.7.

Only selected chlorides are shown in Table 5.2. Chlorides in the group oxidation state are included where these are known, together with some common lower oxidation state species, but a number of other low oxidation state or sub-chlorides such as B_8Cl_8 and related compounds are not shown.

Table 5.2 A structural classification of a selection of the s- and p-block element chlorides. The blocks containing those compounds classified as polymeric are shaded for clarity

1	2		13	14	15	16	17
LiCl	$BeCl_2$		BCl_3	CCl_4	NCl_3		
NaCl	$MgCl_2$		$AlCl_3$	$SiCl_4$	PCl_5	SCl_2	
					PCl_3		
KCl	$CaCl_2$		$GaCl_3$	$GeCl_4$	$AsCl_5$	$SeCl_4$	BrCl
					$AsCl_3$		
RbCl	$SrCl_2$		$InCl_3$	$SnCl_4$	$SbCl_5$	$TeCl_4$	ICl_3
			InCl	$SnCl_2$	$SbCl_3$		
CsCl	$BaCl_2$		$TlCl_3$ TlCl	$PbCl_2$	$BiCl_3$		

ionic   polymeric molecular   polymeric
                     covalent

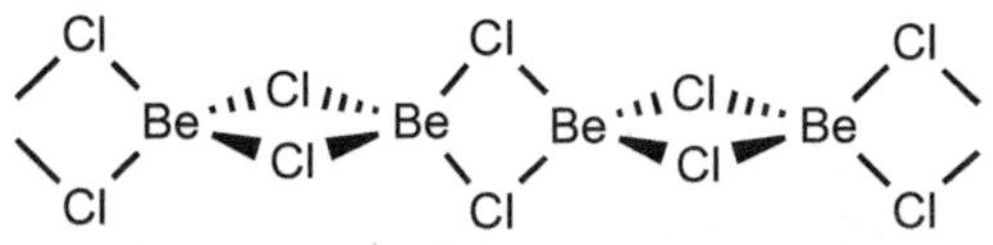

Fig. 5.7 Part of the polymeric structure of $BeCl_2$.

In **Group 13**, $TlCl_3$ can be considered as ionic, but we should also include in this classification the monochlorides of thallium and indium. Both have typically ionic-type structures; TlCl has the CsCl structure (see below) and InCl has a distorted NaCl structure (which is the same as NaF shown in Fig. 5.1(a)). Note that in TlCl the Tl(I) cation is eight-coordinate, whereas in InCl the indium is six-coordinate. These differences undoubtedly reflect the larger size of Tl(I) *vs* In(I), a general point we shall consider below.

The solid-state structures of $AlCl_3$, $GaCl_3$, and $InCl_3$ are probably best described as polymeric. This is certainly the case for $AlCl_3$ which adopts a layer structure with six-coordinate aluminium centres, but for the trichlorides of gallium and indium the description is a little strained in that these compounds exist as dimers in the solid state (Fig. 5.8). In contrast to its heavier congeners, BCl_3 is monomeric probably as a result of appreciable B–Cl π-bonding. We noted in Chapter 4 that π-bond energies are often substantial where first row elements are involved, and this is the likely explanation for the monomeric character of all the boron trihalides. The polymeric nature of the halides of the heavier congeners reflects a tendency for single bond formation, achieved in this instance by halide bridge formation, and the larger size of the elements means they are more able to support a larger coordination number.

In **Group 14**, the tetrachlorides of carbon, silicon, germanium, and tin ($PbCl_4$ is not stable) are all molecular, whereas the dichlorides of tin and lead are polymeric. A view of the polymeric structure of $SnCl_2$, illustrating the three-coordination at tin and the presence of the lone pairs, is shown in Fig. 5.9, and we note again the change from molecular to polymeric resulting from the change in oxidation state of the central element. Recall that a similar transition from one structural type to another was observed for GeF_4 *vs* GeF_2.

All of the chlorides of **Group 15** can be described as molecular, but there is an interesting point to make regarding PCl_5 which is ionic in the solid state, not as a result of the formation of naked phosphorus cations and chloride anions, but due to chloride exchange which affords the salt $[PCl_4]^+ [PCl_6]^-$. A related situation is found for the phosphorus bromide PBr_5 which exists as $[PBr_4]^+ Br^-$. This ionization is not so much a result of particularly polar P–Cl or P–Br bonds, but rather is probably due to the significant lattice energy which results from the formation of an ionic solid, this being considerably larger than any van der Waals interaction between neutral molecules.

In **Groups 16 and 17** the chlorides SCl_2 and BrCl are molecular, whilst $SeCl_4$, $TeCl_4$, and ICl_3 are polymeric. For the latter three compounds, oligomeric is a better description since the former two are tetrameric and ICl_3 exists as the dimer I_2Cl_6 (the structures are shown in Fig. 5.10), but as noted previously, the general descriptor 'polymeric' is used for the reasons given.

For the sake of the arguments being made, the term polymeric is used here to encompass genuinely polymeric structures along with small oligomers (e.g. trimers and tetramers) and dimers. Regardless of the degree of aggregation, the nature of the bonding is similar in all cases.

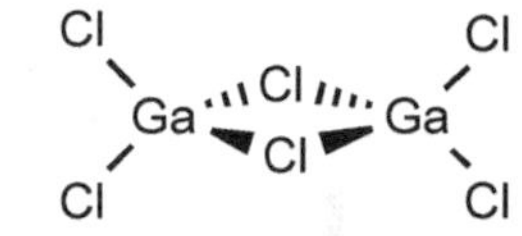

Fig. 5.8 The structure of Ga_2Cl_6.

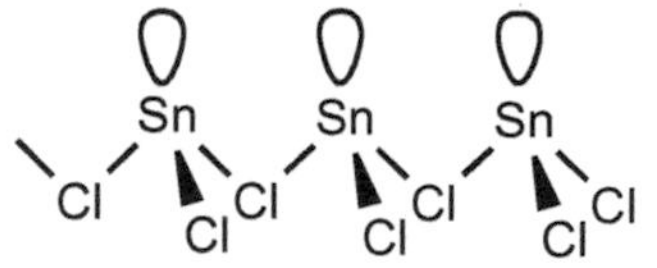

Fig. 5.9 Part of the polymeric structure of $SnCl_2$.

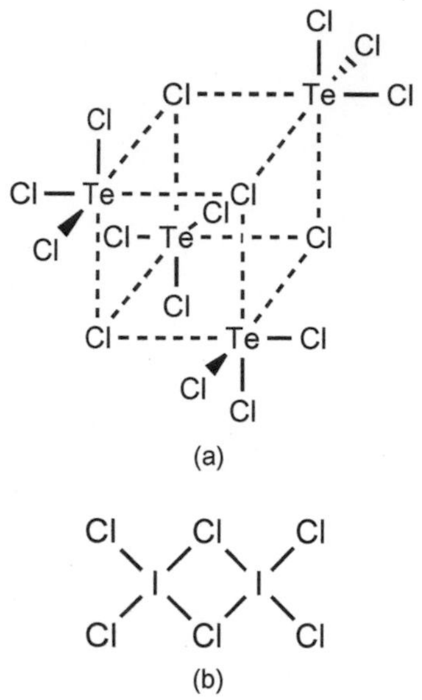

(a)

(b)

Fig. 5.10 The solid-state structures of (a) TeCl$_4$ and (b) ICl$_3$ (I$_2$Cl$_6$).

Other similar equations can also be used, e.g. the Born–Mayer and Kapustinskii equations.

We will not dwell on the structures of the element bromides and iodides, since the same general rules which we have encountered for fluorides and chlorides apply equally well for the heavier congeners. There is a general trend towards less ionic structures, as we might expect, and the number of stable compounds in the group oxidation state becomes progressively less, due partly to the fact that bromide and iodide are more readily oxidized and for other reasons such as halide size discussed in Chapter 4.

Before moving on to look at element oxides, we shall look at ionic structures in a bit more detail and examine some of the general trends observed.

5.2 Ionic structures

This is not the place to go into a lot of detail about the structures and properties of ionic compounds, but it will be useful to briefly consider some general points.

Lattice energies

The first point we shall start with concerns the strength of ionic bonding which we first met in Section 4.4. The strength of the bonding between ions is given by the *lattice energy* which is defined more precisely below, and which can be calculated or measured.

For compounds for which the ionic model is a good representation of the bonding (i.e. a structure in which there is no significant covalent bonding and the ions are treated as hard spheres), lattice energies can be calculated with reasonable accuracy using the Born–Landé equation shown in Eqn. 5.1; here E (usually quoted in kJ mol^{-1}) is the lattice energy ΔH_{Latt} (strictly, lattice enthalpy).

$$E = -\frac{N_A M z^+ z^- e^2}{4\pi\varepsilon_0 r_0}\left(1-\frac{1}{n}\right)$$

(5.1)

The terms in Eqn. 5.1 are as follows: N_A is Avogadro's constant, M is the Madelung constant which relates to the crystal geometry, z^+ and z^- are the charges on the ions, e is the elementary charge, ε_0 is the permittivity of free space, r_0 is the interionic distance (i.e. the sum of the ionic radii, r^+ and r^-), and n is the Born exponent (typically between 5 and 12 and is a measure of repulsive terms), but the important point is that the lattice energy is proportional to the product of the ionic charges and is inversely proportional to the interionic distance such that lattice energies will be largest for structures containing small, highly charged ions.

Lattice energies are defined as the energy (enthalpy change) required to convert a mole of solid into a mole of separated ions in the gas phase, and this can be difficult (if not impossible) to measure directly, since ions may aggregate into small clusters in the gas phase leading to lattice energies being underestimated. Better is to measure them indirectly by means of a **Born–Haber cycle** in which the other terms are amenable to more accurate direct experimental measurement. Such an approach relies on **Hess's Law**, which states that the enthalpy change of a reaction or process is the same regardless of whether it is measured in steps or as a complete process (a consequence of the second law of thermo-dynamics). Examples of Born–Haber cycles can be found in many of the general inorganic texts listed in the bibliography.

Some selected lattice energies are given in Table 5.3 from which a number of trends are apparent. The values for the sodium halides decrease as the size of the halide increases in line with what is to be expected from the Born–Landé equation (i.e. r_0 is increasing). For the Group 2 difluorides, a similar trend is observed on moving down the group, but note how much larger the values are; this is also to be expected from the Born–Landé equation since the cation now has a 2+ charge, i.e. $z^+ = 2$. The same trend is seen for the Group 2 oxides, but note that the values are even larger as the ions are now both doubly charged; $z^+ = z^- = 2$.

Although it is generally the case that the other terms in the Born–Haber cycle (i.e. heats of atomization, ionization energies, electron affinities, and heats of formation) are amenable to more accurate direct experimental measurement, accurate electron affinities are difficult to measure and cannot be measured at all for the O^{2-} ion for which the second electron affinity is endothermic (see later). Some input from theory may still therefore be required.

Table 5.3 Selected lattice energies in kJ mol^{-1}

Compound	Lattice energy	Compound	Lattice energy	Compound	Lattice energy
NaF	930	MgF$_2$	2920	MgO	3900
NaCl	780	CaF$_2$	2600	CaO	3500
NaBr	750	SrF$_2$	2460	SrO	3340
NaI	700	BaF$_2$	2310	BaO	3140

Values in Table 5.3 have been taken from a number of sources and are rounded values where sources differ on the precise value (although they do not generally differ by much). As we have noted for previous Tables, it is general trends in which we are interested, not precise values.

Lattice energies also provide a rationale for a number of physical and chemi-cal properties, and we shall briefly look at some selected examples. Thus, melting and boiling points tend to increase with increasing lattice energy as more energy is required to overcome the interionic forces. For example, compare the melt-ing and boiling points (in °C) of NaF (993, 1695), MgF$_2$ (1263, 2260), and MgO (2582, 3600). These are very high values!

Lattice energies offer a guide to solubility as well. Thus, for an ionic compound to dissolve, the lattice energy must be overcome, which requires that the solvation energies of the dissolved ions (the enthalpy change associated with the binding of solvent molecules to the gas-phase ions) are broadly similar in magnitude to the lattice energies. In the case of water as the solvent, hydration energies can be large enough to be comparable to some of the smaller lattice energies quoted in Table 5.3. Thus, all of the sodium halides are readily soluble in water whereas none of the Group 2 oxides dissolve in water. The lighter Group 2 fluorides are very in-soluble, but solubilities increase down the group; BaF$_2$ is slightly soluble in water.

Finally, we can also rationalize the thermal stability of certain compounds. An example often used is the thermal stability of the Group 2 carbonates, MCO$_3$ (M = Mg, Ca, Sr, Ba). When heated, the carbonates decompose according to Eqn. 5.2, but the temperature required for this decomposition increases sub-stantially as the group is descended. The standard explanation offered is that the lower decomposition temperature for the lighter elements (only 350°C for MgCO$_3$) is a result primarily (ignoring any kinetic factors) of the very high lattice energy for the lighter element oxide, which we can appreciate from the values in the third column of Table 5.3.

$$MCO_3 \rightarrow MO + CO_2 \tag{5.2}$$

Structures

Let us now consider ionic structures themselves. A great many ionic structures can be understood on the basis of the anions constituting an array of close-packed spheres. We met close-packing of spheres in Chapter 3, Section 3.2 with the descriptions of the Group 1 and 2 element structures, so recall that there are two types of close-packed structures, namely hexagonal close-packed (hcp) and cubic close-packed (ccp); the body-centred cubic structure is a more open structure which is not quite close-packed. Any close-packed structure contains holes (or spaces or voids) between the close-packed spheres, these being of two types, either octahedral holes or tetrahedral holes. As the names suggest, anything in an octahedral hole is surrounded by an octahedron of close-packed spheres (i.e. is octahedrally coordinated) and likewise for anything in a tetrahedral hole; it is tetrahedrally coordinated by four spheres. For every sphere in a close-packed structure there is one octahedral hole and two tetrahedral holes. The point of all this is that many ionic structures can be described as comprising a close-packed array of anions with the cations residing in some or all of the octahedral holes or tetrahedral holes (or sometimes a combination of both for more complex structures); occasionally, it is the cations which are considered as close-packed.

Examples are the structure of NaF shown in Fig. 5.1(a), which comprises cubic close-packed fluorides with sodium cations occupying every octahedral hole; this is also the structure adopted by NaCl with cubic close-packed chlorides. The symmetry of the arrangement of the ions in NaF and NaCl is such that it could just as easily be stated that it is the sodium cations which are cubic close-packed with the fluoride or chloride ions occupying the octahedral holes, but it makes more sense to consider (here and more generally for anions) that the halides are close-packed because they are the larger ion (the relevant ionic radii in Å are: Na^+ (0.98), F^- (1.33), Cl^- (1.81)).

The CaF_2 structure (Fig. 5.1(b), the fluorite structure) is an example where it is the cations which are considered close-packed (in this case, cubic) and where the fluorides occupy all the tetrahedral holes, although this is slightly perverse considering the relevant ionic radii (in Å) (Ca^{2+} (1.06), F^- (1.33)). The antifluorite structure is another very common structure and is adopted by, for example, K_2O. Here the oxide anions are close-packed (cubic) with the potassium ions in all the tetrahedral holes.

A third example is the caesium chloride structure shown in Fig. 5.11 in which the chloride anions adopt a non-close-packed body-centred cubic structure with the caesium cations in cubic holes, i.e. surrounded by a cube of eight chlorides.

The structures of numerous solids can be considered in this manner, and much more detail is available in the standard inorganic and solid-state texts listed in the bibliography. A few selected examples (including a couple which contain d-block

Strictly speaking, it is correct to think of atoms in close-packed metallic structures as being genuinely close-packed, i.e. in contact with each other, whereas in ionic structures the ions (usually anions) are not close-packed in such a literal sense (i.e. they are not in direct contact), but the description remains valid and useful in terms of the three-dimensional arrangement.

Note that in a structure of general formula AX, the coordination numbers of both A and X must be equal, e.g. 6 in NaCl. In an AX_2 structure like CaF_2, the coordination number of A is twice that of X; 8 and 4 in this case for Ca and F. In general, for A_nX_m, $CN_X = n/m \times CN_A$, where CN_A and CN_X are the coordination numbers (CN) of A and X respectively.

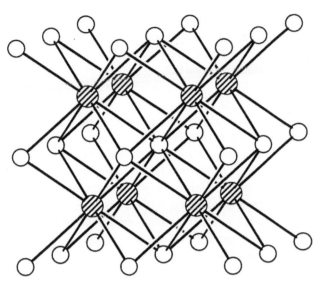

Fig. 5.11 A representation of the structure of caesium chloride, CsCl. The chlorides are the unfilled circles.

elements) which illustrate this approach for hexagonal and cubic close-packing are listed below in Table 5.4. In all cases, it is the anion which is close-packed.

In the perovskite structure, the Ba and O atoms together constitute the cubic close-packed array with the Ti atoms in ¼ of the octahedral holes. An example which does not contain a d-block element is $MgSiO_3$, which is a silicate stable only under high pressure, having octahedrally coordinated silicon centres rather than the more usual tetrahedrally coordinated silicon (this and related minerals are thought to be important in the Earth's mantle where pressures are very high).

Having looked at the structures, the question then arises of what determines which structure is adopted by any particular compound. This is a big topic and the details are beyond the scope of this text (see the bibliography for a number of

The salt $[MeNH_3][PbI_3]$, an important photovoltaic material, also adopts a perovskite structure in which the ammonium cations and iodide anions constitute the close-packed arrangement with the lead cations in ¼ of the octahedral holes.

Table 5.4 Selected ionic structures which illustrate close-packing

Close-packing	Compound	Octahedral holes filled	Tetrahedral holes filled
ccp	NaCl	all	none
ccp	ZnS (zinc blende)	none	½
ccp	K_2O (antifluorite)	none	all
ccp	$MgAl_2O_4$ (spinel)	½ (Mg)	¼ (Al)
ccp	$BaTiO_4$ (perovskite)	¼	none
hcp	ZnS (wurtzite)*	none	½
hcp	PbI_2	½	none
hcp	BiI_3	⅓	none

* BeO is a main group compound with the wurtzite structure.

Pauling proposed a number of rules to explain and rationalize ionic structures and only the first is mentioned here, since an explanation of the others would require more space than is warranted in this text. Solid-state inorganic chemistry references listed in the bibliography have much more detail, particularly Müller (1993).

In structures where only a fraction of the octahedral or tetrahedral holes are occupied, the precise arrangement of occupation varies between different structural types, but it is almost always regular rather than irregular.

general texts which treat this subject in more detail), but we can consider the basics. We cannot easily determine whether the anions will be cubic close-packed or hexagonal close-packed, but a very good guide as to whether cations go in the octahedral holes or the tetrahedral holes is afforded by the ***radius ratio rule***, the radius ratio being that between the cation and the anion given by r_c/r_A. Table 5.5 below illustrates that for particular ranges of radius ratios, certain coordination numbers (hence hole occupancies) are preferred. The reason for this, which is the basis for ***Pauling's first rule***, is that small cations (relative to the anions) will maximize their electrostatic interactions with the anions by going into the smaller holes (i.e. in the smaller tetrahedral rather than the larger octahedral holes). For larger cations, the tetrahedral holes become too small and so occupancy of the octahedral holes is preferred. For even larger cations even the octahedral holes are too small, and the anions are forced apart to give a more open structure such as seen for CsCl. For a more detailed discussion of radius ratios, see Keeler and Wothers (2008).

Table 5.5 Radius ratios and hole occupancy

Radius ratio	Coordination number	Class of hole filled	Example
0.225–0.414	4	tetrahedral	BeO
0.414–0.732	6	octahedral	NaCl
0.732–1.000	8	cubic	CsCl

The description of ionic structures presented here (close-packed spheres, octahedral and tetrahedral holes etc.) is not the only available description. It is often useful to think of an anion coordination polyhedron around cation centres (octahedral and tetrahedral being the most common) and then view the structures as based on polyhedra linked through vertex-, edge-, or face-sharing. Some of Pauling's other rules deal with this type of description.

Radius ratios are not a faultless way of predicting or rationalizing all structures by any means, particularly for more complex formulae, but they work reasonably well and are certainly a useful guide to the type of structure adopted. They work best for compounds for which the ionic model of electrostatically interacting charged spheres is a good description of the bonding, and that is generally the case where $\Delta\chi$ is large. For smaller values of $\Delta\chi$, where there is some degree of covalency, exceptions are more likely. For example, some AX structures for which the radius ratio would predict occupancy of the octahedral holes for A adopt structures where A is in a tetrahedral hole, tetrahedral four-coordination becoming favoured over octahedral six-coordination as covalent bonding becomes more important.

5.3 Element oxides

A classification of the s- and p-block element oxides according to whether they are ionic, polymeric, or molecular covalent is presented in Table 5.6, and the similarity to Tables 5.1 and 5.2, is obvious.

As we would expect, the left-hand **Group 1** and **Group 2** element oxides are predominantly ionic due to the large electronegativity difference with oxygen, but we will not comment on these compounds or review the structures in any detail except to point out that the structures are typical of ionic materials referred to in the previous section. We should be aware, however, that there are occasions when the classification, particularly that between ionic and polymeric,

Table 5.6 A structural classification of the s- and p-block element oxides. The block containing those compounds classified as polymeric is shaded for clarity

1	2	13	14	15	16	17	18
Li_2O	BeO	B_2O_3	CO	NO	O_2	F_2O	
			CO_2	N_2O	O_3	F_2O_2	
				N_2O_3			
				N_2O_4			
				N_2O_5			
Na_2O	MgO	Al_2O_3	SiO_2	P_4O_6	SO_2	Cl_2O	
				P_4O_{10}	SO_3	ClO_2	
						Cl_2O_7	
K_2O	CaO	Ga_2O_3	GeO_2	As_4O_6	SeO_2	Br_2O	
				As_2O_5	SeO_3	BrO_2	
Rb_2O	SrO	In_2O_3	SnO_2	Sb_4O_6	TeO_2	I_2O_4	XeO_3
			SnO	Sb_2O_5	TeO_3	I_2O_5	XeO_4
						I_4O_9	
	BaO	Tl_2O_3	PbO_2	Bi_2O_3	PoO_2		
		Tl_2O	PbO				

ionic	polymeric	molecular covalent

Only a selection of element oxides is shown in Table 5.6, and not shown at all are the superoxides and peroxides of Group 1, i.e. EO_2 and E_2O_2 which contain the O_2^- and O_2^{2-} ions respectively. Interestingly, lithium superoxide, LiO_2, is not stable except at very low temperatures, unlike the oxide Li_2O and peroxide, Li_2O_2, presumably because lattice energy considerations favour the small Li^+ in association with a 2– ion much more than with a 1– ion.

Polonium dioxide is exceptional in that it best described as ionic and so is drawn in a box within the section for polymeric compounds.

is somewhat arbitrary. A further point we should recognize concerns the oxide anion O^{2-}. Although the first electron affinity of oxygen to form O^- is exothermic (Table 2.3), the second electron affinity, i.e. O^- to O^{2-}, is strongly endothermic. The fact that the oxide anion O^{2-} is present in ionic oxides is a function of the large lattice energy associated with these structures (see Section 5.2).

In **Group 13**, all structures can be described as polymeric. Again, we will not look at the structures in detail, although it is worth mentioning that there are a number of different structural types, most of which involve octahedral coordination of the Group 13 atoms and four-coordinate oxygens.

In **Group 14** the trends are more varied and informative. Both of the common oxides of carbon, CO and CO_2, are molecular covalent species which are gases under ambient conditions (Fig. 5.12), the latter in striking contrast to SiO_2 which is a polymeric, hard, and refractory solid; part of the β-crystobalite structure (one of the polymorphs of SiO_2) is shown in Fig. 5.13.

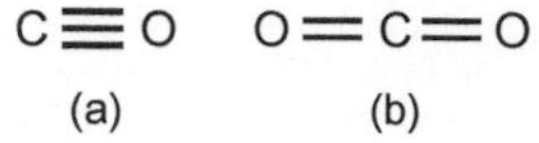

$$C \equiv O \qquad O = C = O$$

(a)   (b)

Fig. 5.12 The structures of (a) CO and (b) CO_2.

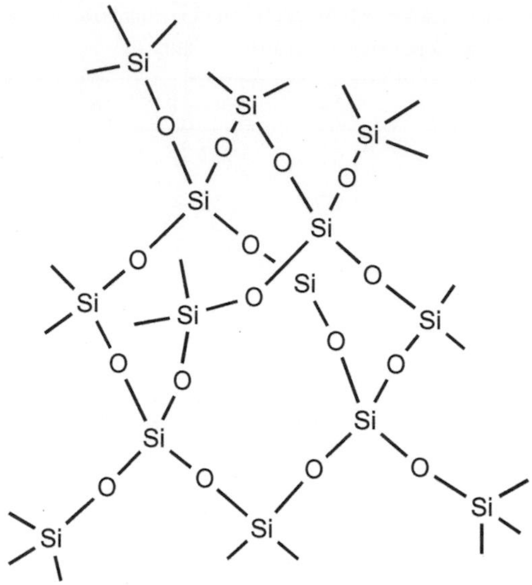

Fig. 5.13 Part of the structure of the β-crystobalite form of SiO_2.

The structural differences between CO_2 and SiO_2 are consistent with the electronegativity difference between C and O *vs* Si and O, the greater difference being between silicon and oxygen, but this represents only part of the explanation. CO_2 is a linear molecule containing two-coordinate carbon in which strong C–O π-bonding plays an important rôle. On the basis of the discussion in Section 4.4 on multiple bonds, this is not surprising, and we would expect that whereas for carbon, a C=O double bond is stronger than two C–O single bonds (2 × 335 *vs* 715 kJ mol⁻¹, Table 4.9), the opposite will be true for silicon–oxygen bonds such that silicon will prefer to form two Si–O single bonds rather than one Si=O double bond (2 × 420 *vs* 590 kJ mol⁻¹, Table 4.9). An isolated SiO_2 molecule would therefore be expected to polymerize to give a macromolecular solid with four-coordinate silicon and two-coordinate oxygens in which all Si–O bonds are single; this is exactly what is found in the majority of the polymorphs of SiO_2. We should also consider the relative sizes of the elements; thus, the larger silicon atom is better able to accommodate the larger coordination number, i.e. four *vs* two (six-coordinate silicon is also known, as previously mentioned for $MgSiO_3$).

Germanium dioxide has a structural chemistry that is very similar to that of silicon dioxide, which is not surprising in view of the similar size and electronegativity of the two elements. In contrast, the structures of tin dioxide and lead dioxide (one polymorph) are of the rutile type (one of the polymorphs of TiO_2) in which the metal atoms are octahedrally coordinated by oxygen atoms. Other polymorphs of PbO_2 also feature octahedrally coordinated lead, and in all cases this increased coordination number of Sn and Pb *vs* Ge and Si can be traced to the larger size of the atoms.

The structure of tin monoxide (Fig. 5.14) consists of planar arrays of oxygen atoms with squares of oxygens alternately capped by tin atoms; the structure of

Fig. 5.14 Part of the solid-state structure of SnO, showing the coordination around the tin atom.

lead monoxide is isomorphous. The lower coordination number of the tin atom (*vs* that in SnO_2) results from the presence of the lone pair of electrons associated with the tin(II) centre (*cf* the halides) which resides at the apex of the square pyramid defined by the tin and four oxygen atoms. This structure is sometimes classified as ionic, but the layer or two-dimensional aspect is better described as polymeric, since ionic structures are generally more uniformly three-dimensional in nature.

The structures of the oxides of the **Group 15** elements follow a rather similar pattern to the oxides of Group 14. Thus, all the oxides of nitrogen are molecular covalent, feature strong N–O π-bonding and have coordination numbers around the nitrogen atoms which do not exceed three. The structures are shown in Fig. 5.15.

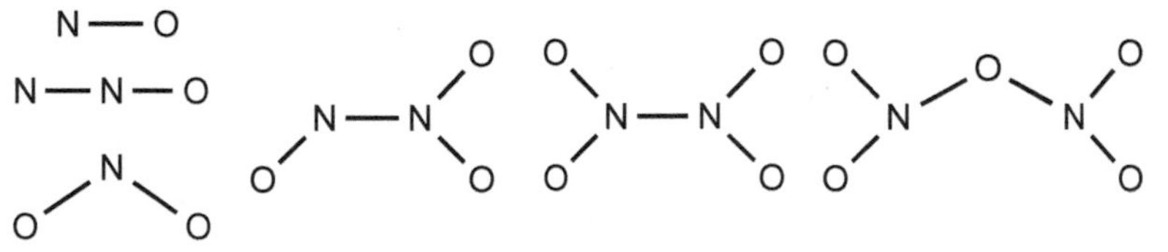

No attempt is made to indicate bond order or multiplicity in Figs. 5.15–5.17 or 5.19.

Fig. 5.15 Structures of the oxides of nitrogen: NO, N_2O, NO_2, N_2O_3, N_2O_4, N_2O_5.

In contrast, none of the molecular nitrogen oxides has any counterpart in phosphorus chemistry (except as high-temperature species). The combination of increased electronegativity difference, much weaker P–O π-bonding, and the larger size of the phosphorus atom all favour polymeric forms exactly as seen in Group 14. The common oxides of phosphorus are P_4O_6 and P_4O_{10}, the structures of which are shown in Fig. 5.16. The term 'oligomeric' might be more appropriate, although only one phase of P_4O_{10} contains discrete molecules; two others are also known which comprise two- and three-dimensional macromolecular structures although still based on PO_4 tetrahedra sharing three vertices.

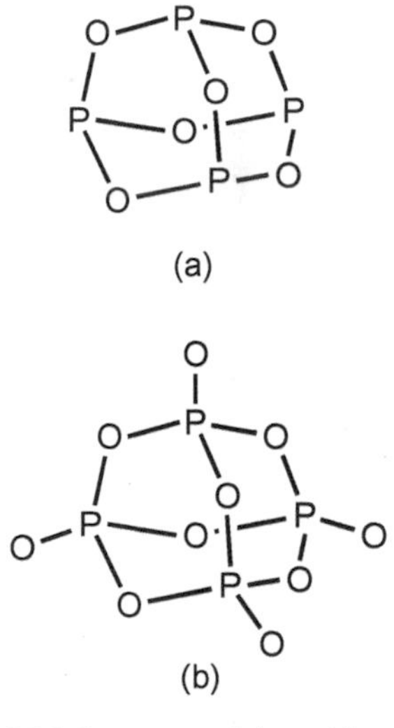

The oxides E_2O_3 of arsenic and antimony both have phases containing molecular E_4O_6 units analogous to P_4O_6, but also exist in polymeric forms consisting of edge-sharing EO_3 pyramids. Bi_2O_3, in contrast, is exclusively polymeric in which the larger bismuth atom attains a coordination number of five in a variety of polymorphs. The pentavalent oxides of arsenic, antimony, and bismuth are less well characterized but are certainly polymeric and contain the elements octahedrally coordinated.

The oxides of the **Group 16** elements provide another interesting series which illustrates the effects of element size, electronegativity differences, and multiple bonding. The elemental forms of oxygen (i.e. the 'oxides' of oxygen) are O_2 and O_3, for both of which multiple bonding is important with concomitant low coordination numbers. With sulfur, the similar electronegativity to oxygen favours the formation of molecular covalent species, but the relatively small size of the sulfur atom means that π-bonding is still fairly strong, since orbital overlap is good. As a consequence, the most important oxide of sulfur, SO_2, is not only molecular covalent but also contains S–O π-bonding. The other main

Fig. 5.16 Structures of the oxides of phosphorus, (a) P_4O_6 and (b) P_4O_{10}.

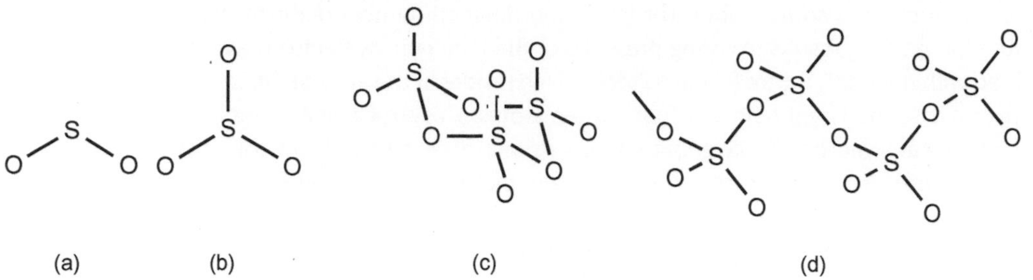

Fig. 5.17 The structures of (a) SO_2, (b) monomeric SO_3, (c) cyclotrimeric γ-SO_3, and (d) helical β-SO_3.

oxide is SO_3, which is molecular in the gas phase but which exists as a cyclotrimer in the solid γ-phase or as a helical polymer in the β-phase; these structures are shown in Fig. 5.17.

It is worth commenting on why SO_2 is exclusively monomeric whereas SO_3, in the solid state, is polymeric, since this would appear to be at odds with what we observed for the germanium fluorides GeF_2 and GeF_4, i.e. that the lower oxidation states are the ones which tend to polymerize. For SO_3, a possible explanation is that the presence of three electron-withdrawing oxygens around the sulfur in SO_3 induces a sufficiently large δ^+ charge at the sulfur centre to promote oligomerization and that the coordination number is readily increased from three to four. In effect, the sulfur in SO_3 is coordinatively unsaturated, whereas the germanium centre in GeF_4 is not. In this regard we may note that trigonal planar, three-coordination is unusual for sulfur. The reason why GeF_2 polymerizes whereas SO_2 does not is probably a consequence of the much more polar bonds in the former and the S–O π-bonding in the latter. We may note also that the ionic species SiO_3^{2-} and PO_3^{-}, which are isoelectronic with SO_3, are exclusively polymeric in the solid state in contrast to their lighter congeners CO_3^{2-} and NO_3^{-} which are always monomeric; size effects are important here, together with the strength of multiple bonding when first-row elements are involved.

Selenium dioxide, SeO_2, in contrast to SO_2, is polymeric with a linear zig-zag structure shown in Fig. 5.18(a). Since the electronegativity difference between S and Se is small, weaker π-bonding and the larger size of the selenium atom are the likely reasons for this difference.

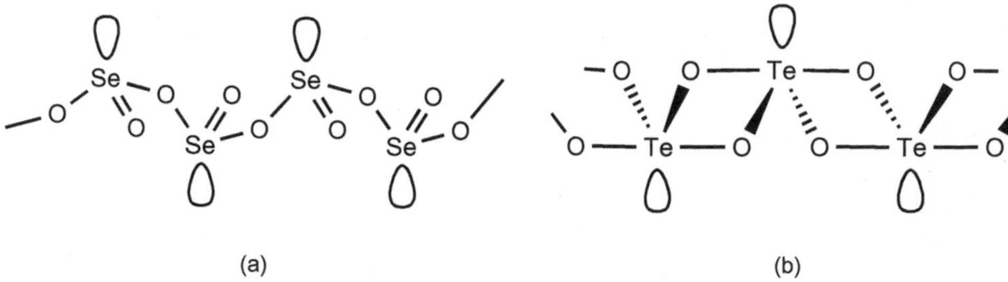

Isoelectronic means having the same number of electrons, often simply the same number of valence electrons.

Fig. 5.18 Part of the structure of (a) SeO_2 and (b) TeO_2.

We should further note that GeF_2, SO_2, and SeO_2 are all isoelectronic with respect to their valence electrons. The similarity between the structures of GeF_2 and SeO_2 is obvious from a comparison of Figs. 5.4(b) and 5.18(a), and the fact that SO_2 is monomeric is therefore all the more noticeable. That SO_2 has the smallest $\Delta\chi$ of the three examples is significant, together with the fact, as noted previously, that π-bonding would be expected to be strongest between S and O. Note that whilst conventional electron counting procedures would assign GeF_2 a vacant orbital (as a result of having Ge–F single bonds), this would not be the case with SO_2, which is represented with S=O double bonds to indicate the importance of π-bonding.

Tellurium dioxide is also polymeric, although the coordination number of the tellurium atoms in the various polymorphs which are known is four, and all Te–O bonds are single as shown in Fig. 5.18(b). Polonium dioxide has the fluorite structure in which the Po atoms are eight-coordinate and the structure can be described as ionic. Note that the dioxides of Group 16 clearly illustrate the trend to larger coordination numbers as the group is descended; two for SO_2, three for SeO_2, four for TeO_2, and eight for PoO_2, and the trend from covalent molecules through polymeric structures to an ionic compound.

The higher oxidation state oxides are SeO_3 and TeO_3; the former is a cyclotetramer (in contrast to cyclotrimeric SO_3) and the latter consists of TeO_6 octahedra sharing all vertices to give a three-dimensional lattice.

In **Group 17**, the oxides of fluorine, chlorine, and bromine are all molecular covalent, and π-bonding is important to some extent. Selected structures of some chlorine oxides are shown in Fig. 5.19(a)-(c). The most stable oxide of iodine is I_2O_5, the molecular unit of which is shown in Fig. 5.19(d), although in the solid state, molecules are associated to give a more polymeric structure consistent with the greater electronegativity difference between I and O.

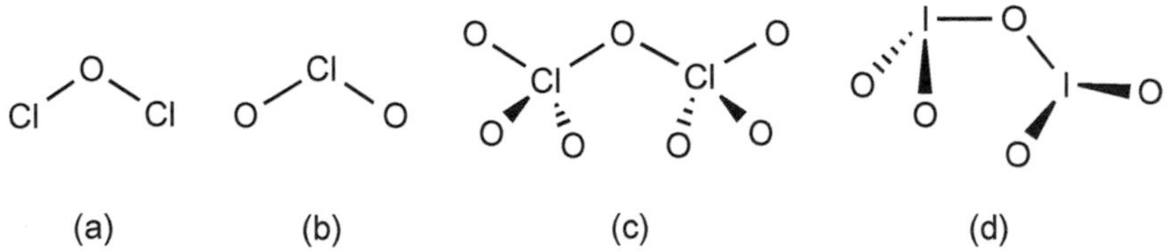

(a)   (b)   (c)   (d)

Fig. 5.19 Some oxides of chlorine and iodine: (a) Cl_2O, (b) ClO_2, (c) Cl_2O_7, and (d) I_2O_5.

In **Group 18** the only stable neutral oxides are those of xenon, XeO_3 and XeO_4, which are both molecular with trigonal pyramidal and tetrahedral geometries respectively.

5.4 Element hydrides

A structural classification of the s- and p-block hydrides is shown in Table 5.7, and the general patterns which we have seen for the halides and oxides are also apparent here. With the exception of BeH_2, the hydrides of **Group 1** and **Group 2** all adopt typically ionic structures and also have lattice ener-

Table 5.7 A structural classification of the s- and p-block element hydrides. The block containing those compounds classified as polymeric is shaded for clarity

1	2	13	14	15	16	17
LiH	BeH_2	BH_3	CH_4	NH_3	H_2O	HF
NaH	MgH_2	AlH_3	SiH_4	PH_3	H_2S	HCl
KH	CaH_2	GaH_3	GeH_4	AsH_3	H_2Se	HBr
RbH	SrH_2		SnH_4	SbH_3	H_2Te	HI
CsH	BaH_2		PbH_4	BiH_3		
	ionic	polymeric		molecular	covalent	

The ionic radius of the hydride anion, H⁻, is rather variable depending on its environment but is usually placed between bromide and iodide. Thus, it is quite large, and this reflects the relative ineffectiveness of a single proton to hold onto two electrons and the resulting dominance of interelectron repulsions. It is therefore not surprising that the hydride anion is easily polarized.

gies broadly consistent with predominantly ionic bonding. The exceptions to this are the hydrides of the lighter elements such as LiH and MgH_2 which have some degree of covalency, although this is to be expected in view of the high polarizing power of these small cations and the high polarizability of the hydride anion.

Beryllium hydride, BeH_2, is best described as polymeric and adopts a linear structure in the solid state with bridging hydrides (similar to $BeCl_2$); the extreme polarizing ability of the hypothetical Be^{2+} ion would make the ionic model inappropriate here in contrast to the hydrides of the heavier Group 2 metals.

In **Group 13** the hydrides are also polymeric, although we must be a little careful here in the use of electronegativity arguments (see AlF_3 earlier). Boron forms a very large number of binary hydrides which can only be satisfactorily described in terms of delocalized, multicentre bonding; we shall just look at the simplest hydride, BH_3. Borane is a gas under ambient conditions and exists as a dimer diborane, B_2H_6, which has the hydride-bridged structure shown in Fig. 5.20.

Descriptions of the bonding in the B–H–B bridges in B_2H_6 and the higher boron hydrides can be found in most standard inorganic texts listed in the bibliography.

We can stretch our use of the word 'polymer' to a dimer if we wish (we did for Ga_2Cl_6 and I_2Cl_6), but it is important to realize that the dimerization is mostly a consequence of the presence of a vacant orbital on the boron centre rather than any difference in electronegativity between the elements. In fact, boron and hydrogen have very similar electronegativities (2.04 and 2.02 respectively on the Pauling scale) and therefore essentially non-polar B–H bonds. A similar situation is found for gallane (Ga_2H_6), which is isostructural with diborane, and for similar reasons; the Pauling electronegativity of gallium is 1.81.

Fig. 5.20 The structure of B_2H_6.

Binary, neutral hydrides of indium and thallium are not well characterized as a result of the extreme weakness of the covalent In–H and Tl–H bonds; a factor in common with the hydrides of all the heavier p-block elements. Aluminium hydride is a volatile solid under ambient conditions which can certainly be described as polymeric since it exists as a three-dimensional, hydrogen-bridged structure. This undoubtedly does reflect the electronegativity difference between the elements which is large enough for a polymeric structure to be expected, although the vacant aluminium 3p orbitals are also important.

AlH_3 has certain properties in common with some of the s-block hydrides such that an ionic description is sometimes preferred. Indeed, the α-form is isostructural with AlF_3.

The hydrides of **Group 14**, and indeed of all subsequent groups, are all molecular covalent. Methane, CH_4, is the simplest hydride of carbon but there are an essentially limitless number of higher hydrocarbons in which catenated carbon–carbon bonds are present. Silane, SiH_4, and germane, GeH_4, are the prototypical hydrides of silicon and germanium, although the higher catenated congeners are generally less stable than for carbon, this being particularly the case for germanium. The tin and lead hydrides, SnH_4 and PbH_4, are considerably less stable, which again reflects the weaker heavier element–hydrogen bonds; catenated compounds are particularly unstable and are unknown for lead. It is also worth noting that tin(II) and lead(II) hydrides are unknown, particularly in view of the importance of this oxidation state in the chemistry of these elements, especially lead. This is analogous to the situation noted in Chapter 4 for the lead alkyls. Part of the reason for the instability of the heavier p-block hydrides is the poor overlap between the orbitals of these elements and the 1s orbital of hydrogen. For tin(IV) and lead(IV) there is sufficient overlap for marginal stability, but this is not the case for the larger tin(II) and lead(II) with bigger and more diffuse orbitals.

In **Group 15**, the trihydride, EH_3, is known for all the elements with stability decreasing down the group for reasons already discussed; only for nitrogen and phosphorus are higher hydrides involving catenation known such as N_2H_4 (hydrazine) and P_2H_4. Conspicuous by their absence are any stable hydrides of the higher +5 oxidation state which is probably partly due to the fact that hydride is fairly readily oxidized and is therefore not compatible with elements in high formal oxidation states. Another important additional factor, however, is that the weaker element–hydrogen bonds which would be expected for a pentavalent hydride would probably result in the reaction shown in Eqn. 5.4 being quite exothermic, being driven, in part, by the strong H–H bond.

$$EH_5 \rightarrow EH_3 + H_2 \qquad (5.4)$$

We need to be a little careful with this argument, however. For Groups 13 and 14 we have highlighted the fact that it is the higher oxidation state hydrides that are stable rather than the lower oxidation state ones. This is the case also for the alkyls for the reasons given in Chapter 4, Section 4.4. Nevertheless, for whatever reason, matters are a little different in Group 15, since for all elements, compounds of the type ER_3 are known (R = alkyl or aryl) but as we have also commented on earlier, $BiMe_5$ and $BiAr_5$ have also been characterized.

The **Group 16** hydrides are EH_2 (now usually written as H_2E) and again stability decreases down the group; H_2Po is not stable under ambient conditions. The only stable catenated hydride in this group is hydrogen peroxide, H_2O_2, and hydrides of the higher +4 and +6 oxidation states are unknown as in Group 15.

Group 17 is characterized by the hydrogen halides, HE (usually HX), with no higher oxidation state hydrides, and there are no stable hydrides of the **Group 18** elements.

Downs and Pulham have argued that whilst the weakness of the E–H bonds of the heavier p-block hydrides is a factor regarding their stability, many of the s-block hydrides have weaker E–H bonds. Also important is the strength of the bridging E–H–E interactions which are strong for the s-block hydrides, since they are essentially ionic in nature, but weaker in Group 13/14 where they are more covalent and decrease in strength down the group.

As with the alkyls (see Eqn. 4.6), the disproportionation reaction shown in Eqn. 5.3 is calculated to be exothermic and decomposition generally to the Group 14 element and H_2 will be favoured by the strength of the H–H bond in H_2. Most of the heavier p-block hydrides have endothermic heats of formation.

$$2\ Pb(II)H_2 \rightarrow Pb(0) + Pb(IV)H_4 \qquad (5.3)$$

Strictly speaking, we should refer to a hypothetical NH_5 as pentavalent rather than as N(V) since nitrogen is more electronegative than hydrogen.

Much of the instability of element hydrides discussed in this section may be traced, in part, to the strength of the H–H bond and its influence on the thermodynamics of compound decomposition as noted previously.

5.5 Physical properties

We shall finish this chapter with a brief look at some other general trends in the properties of compounds and start with the physical nature of the element hydrides of Groups 14 and 16.

As shown in Table 5.5, all these hydrides are molecular covalent species, but the trends in melting and boiling points are worthy of comment. In Group 14 we see a progression to higher melting and boiling points as the group is descended resulting from the increasing van der Waals forces between the molecules. This is also largely the case for the Group 16 hydrides but with the obvious exception of H_2O. The explanation for this feature is the extensive intermolecular hydrogen bonding between oxygen lone pairs and hydrogen, this being that much greater for the lighter element oxygen due to its high electronegativity. The values for all compounds are shown in Fig. 5.21. The observed trend for the Group 15 elements is similar to that seen for Group 16 as a result of the importance of hydrogen bonding in NH_3.

An illustration of a hydrogen bonding interaction between two water molecules is shown in Fig. 5.22. Recall that $\Delta\chi$ for O and H will polarize the O–H bond as $(\delta^-)O–H(\delta^+)$, which gives rise to the electrostatic interaction that characterizes intermolecular hydrogen bonding as indicated by the dotted line in Fig. 5.22. This accounts for why hydrogen bonding is particularly important for compounds containing O–H and N–H bonds (and certainly for HF) but much less so for the less electronegative elements in those groups (and hence less polar E–H bonds); since lone pairs are not a feature of carbon chemistry, the issue of hydrogen bonding does not arise and the properties of CH_4 are in no way anomalous. Whilst an electrostatic explanation for hydrogen bonding is

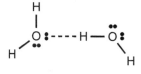

Fig. 5.22 An illustration of the intermolecular hydrogen bonding interaction between two water molecules.

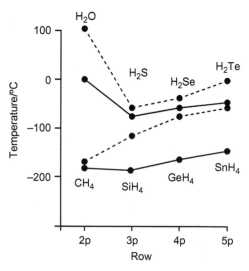

Fig. 5.21 Boiling points (dashed lines) and melting points (unbroken lines) for the hydrides of the Group 14 elements (lower two lines) and the Group 16 elements (upper two lines).

standard, a covalent model involving, for example, a three-centre, four-electron O–H–O interaction, offers some insight into the observed directionality associated with hydrogen bonding.

Another example of the increasing strength of van der Waals interactions on progressing to heavier elements is seen in the physical properties of the carbon tetrahalides, CX_4 (X = F, Cl, Br, I), as shown in Table 5.8.

Table 5.8 Physical properties of the carbon tetrahalides, CX_4 (X = F, Cl, Br, I)

Compound	Phase at RTP*	Colour
CF_4	Gas	Colourless
CCl_4	Liquid	Colourless
CBr_4	Solid	Pale yellow
CI_4	Solid	Dark red/violet

* RTP = room temperature and pressure

Thus, the trend is from gas to liquid to solid as the van der Waals forces between molecules increase as a direct result of the larger number of electrons as we go from F to I. This is quite general; molecular fluorides are often gases or volatile liquids, whereas molecular iodides are generally solids.

The other point to note concerns colour. There is clearly a trend from colourless to coloured as we go from CF_4 to CI_4, and this arises from a decrease in the HOMO-LUMO gap as we move from C–F to C–I bonds. The important transition which gives rise to absorption of radiation in this case is the so-called n→σ* transition, where n denotes a lone pair, i.e. an electron is being promoted from a halogen lone pair (the HOMO) to a C–X σ*-orbital (the LUMO). Overlap is good between C and F so the HOMO-LUMO gap is large and the transition and hence absorption is in the UV so the compound appears colourless. For the larger I, however, with more diffuse orbitals which result in poorer overlap, the HOMO-LUMO gap is small such that transitions now move into the visible resulting in absorption of visible light and hence coloured compounds. This is also quite general; fluorides are often colourless whereas molecular iodides are often strongly coloured. We saw and noted something rather similar for the halogens themselves in Chapter 3.

Remember that van der Waals forces arise from fluctuating, short-lived charge asymmetries in the molecular electron density, the more electrons, the more asymmetry so the stronger the intermolecular forces, hence higher melting and boiling points.

Exercises

Account for the following observations:

1. Boron trichloride, gallium trichloride, and thallium trichloride are, respectively, covalent and monomeric, covalent and dimeric, and ionic.

2. $SnCl_2$ is polymeric whereas $SnCl_4$ is monomeric.

3. CO_2 is molecular, SiO_2 is polymeric, and SnO_2 is ionic, and the coordination numbers around the respective Group 14 element centre increase from 2 to 4 to 6.

4. The structure of sulfur dioxide comprises isolated SO_2 molecules, whereas selenium dioxide and tellurium dioxide adopt polymeric structures with one and two bridging oxygen atoms respectively for each Group 16 atom.

5. CCl_4 is inert to water under all but extreme conditions, whereas $SiCl_4$ is readily hydrolyzed.

6. With regard to the Group 15 trihalides, account for why NF_3 is a colourless gas whereas BiI_3 is a deep orange solid.

6 Acids and bases

6.1 Definitions

There are many definitions of acids and bases, some more specific than others, but this is not the place to cover the topic in great detail. However, some simple definitions of acidity and basicity will be useful at this point before moving on to look at the various trends which occur primarily in the p-block.

Lewis acidity

The broadest definition of acids and bases is the **Lewis** definition. In this scheme a **Lewis acid** is an electron-pair acceptor whilst a **Lewis base** is an electron-pair donor. Two examples will serve as illustrations. In the adduct formed between boron trifluoride, BF_3, and trimethylamine, NMe_3, the boron centre acts as a Lewis acid whilst the nitrogen centre acts as a Lewis base. The resulting N–B bond is called a **coordinate bond** or **dative covalent bond** and is represented by an arrow as shown in Fig. 6.1. The nitrogen is able to act as a base since it has a non-bonded or lone pair of electrons, and the boron centre is acidic because it possesses a vacant p orbital which can accept an electron pair.

As a second example we can look at the adduct formed between SiF_4 (the acid) and pyridine (py, the base) which results in the complex $[SiF_4(py)_2]$ (Fig. 6.2). The nature of the vacant orbital on the silicon centre is something to which we shall return in Section 6.3.

Brønsted–Lowry acidity

The **Brønsted–Lowry** concept is probably the second most commonly employed description of acids and bases, although in aqueous solution it is by far the most widely used. In this scheme an acid is defined as a proton (H^+) donor and a base as a proton acceptor. We can illustrate this with reference to the familiar phrase 'acid plus base gives salt plus water', an example of which is given in Eqn. 6.1. Here, hydrochloric acid, HCl, is the acid, sodium hydroxide the base, and sodium chloride the salt or neutralization product. This definition can readily be extended to other protonic solvent systems such as sulfuric acid and liquid

Fig. 6.1 The structure of $Me_3N{\rightarrow}BF_3$.

The use of an arrow to represent a dative bond is commonplace, but we should recognize that an alternative representation is as a single bond with formal charges (such that each atom obeys the octet rule), i.e. $Me_3N^+–B^-F_3$.

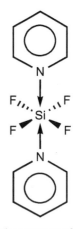

Fig. 6.2 The structure of $[SiF_4(py)_2]$.

ammonia. Moreover, the concept of conjugate acids and conjugate bases is also useful, as indicated in Eqn. 6.2.

$$HCl(aq) + NaOH(aq) \rightarrow NaCl(aq) + H_2O(l) \tag{6.1}$$

$$HS^-(aq) + H_2O(l) \rightarrow S^{2-}(aq) + H_3O^+(aq) \tag{6.2}$$

Thus, S^{2-} is the **conjugate base** of the acid HS^- and H_3O^+ is the **conjugate acid** of the base H_2O. Note that HS^- can act as an acid or as a base, depending on the conditions or pH. Under acidic conditions it acts as a base and is protonated to give H_2S, whilst under basic conditions it acts as an acid and is deprotonated to give S^{2-}. Note that the concept of Lewis acidity and basicity incorporates the Brønsted–Lowry approach as a special case.

In aqueous solution, a proton does not exist in isolation; it is certainly hydrated and is generally written as H_3O^+ (the hydronium ion), although this is also an oversimplification. Larger aggregates such as $H_5O_2^+$ are undoubtedly present as well.

6.2 Element oxides and hydroxides

One important aspect of the element oxides, discussed in Chapter 5, and of the closely related hydroxides, concerns their acidic or basic properties and the relationship between these properties and the position of the element in the periodic table. Fig. 6.3 illustrates this point and we will consider it in detail shortly, but first we should understand what is meant by an oxide (or hydroxide) being described as acidic or basic, and also what is meant by the term **amphoteric** which is encountered in this context.

An acidic oxide, in aqueous solution, is one which combines with a water molecule and then releases one or more protons with concomitant formation of an oxo-anion. This is illustrated in Eqn. 6.3 for sulfur trioxide, SO_3. Thus, addition of SO_3 to water forms sulfuric acid, H_2SO_4, which can lose one proton to form bisulfate, HSO_4^-, or two protons to form sulfate, SO_4^{2-}. All of the elements in bold face in Fig. 6.3 have acidic oxides which therefore react with water to give acidic solutions (because they contain H^+) and oxo-element anions.

Note that SO_3 is also acidic according to the Lewis criteria as exemplified by the following equation, which shows the addition of oxide (a base) to form sulfate:

$$SO_3 + O^{2-} \rightarrow SO_4^{2-}$$

$$SO_3(aq) + H_2O(l) \rightarrow H_2SO_4(aq) \rightarrow H^+(aq) + HSO_4^-(aq) \rightarrow$$
$$2H^+(aq) + SO_4^{2-}(aq) \tag{6.3}$$

An alternative description is that an acidic oxide reacts with an aqueous base, i.e. it acts as a hydroxide acceptor, to give an oxo-anion (or hydroxo-anion) which is then a source of H^+; using the same system as above, we can represent this as shown in Eqn. 6.4.

$$SO_3(aq) + OH^-(aq) \rightarrow HSO_4^-(aq) \rightarrow H^+(aq) + SO_4^{2-}(aq) \tag{6.4}$$

In some cases, hydroxides are acidic because they act solely as hydroxide acceptors. An example of this situation is found with boron trihydroxide or orthoboric acid, $B(OH)_3$ or H_3BO_3. The former formula and name is more appropriate since $B(OH)_3$ is not itself a source of protons in aqueous solution but acts instead as a hydroxide acceptor affording the borate anion $[B(OH)_4]^-$ and thereby lowering the pH according to Eqn. 6.5.

$$B(OH)_3 + 2\ H_2O \rightarrow \left[B(OH)_4\right]^- + H_3O^+ \qquad (6.5)$$

While considering acids, an interesting comparison can be made between the dihydrates of H_2SO_4 and H_2TeO_4. In the former case this is a hydrogen-bonded adduct of molecular H_2SO_4 (represented as $H_2SO_4.2H_2O$) whereas in the latter case, H_2O adds across the formal Te=O bonds to give H_6TeO_6 (or $Te(OH)_6$) which is an octahedral hexahydroxide of Te. The difference between S and Te can be traced to three factors: (i) the larger Te can more readily support a larger coordination number (6 vs 4), (ii) Te prefers two Te–O single bonds as opposed to one Te=O double bond in contrast to S, and (iii) the initial Te=O bond is more polar than the S=O bond, as a result of the greater $\Delta\chi$ for Te/O vs S/O, thereby favouring the addition of H_2O.

A basic oxide reacts with water and dissociates to give an element cation and hydroxide ions as illustrated for calcium oxide in Eqn. 6.6.

$$CaO(s) + H_2O(l) \rightarrow Ca^{2+}(aq) + 2OH^-(aq) \qquad (6.6)$$

An element hydroxide similarly dissociates in water to give an element cation and hydroxide ions, though we should note that the distinction between oxides and hydroxides in aqueous solution is often more formal than real. Thus, addition of H_2O to CaO will initially afford $Ca(OH)_2$ which subsequently dissociates. The elements at the lower left in Fig. 6.3 (in normal face script) all form basic oxides and hydroxides.

As with the acids above, we can employ an alternative description wherein a basic oxide reacts with aqueous acid to form an element cation and water as shown in Eqn. 6.7.

$$CaO(s) + 2H^+(aq) \rightarrow Ca^{2+}(aq) + H_2O(l) \qquad (6.7)$$

A similar contrast is seen in Group 17 chemistry. Thus, in acidic solution, IO_4^- is protonated to form $[I(OH)_6]^+$, the corresponding perchlorate anion, ClO_4^-, showing no such tendency.

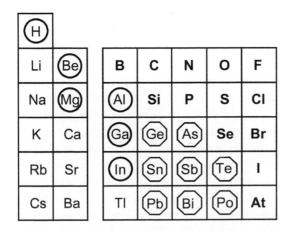

Fig. 6.3 The s- and p-block elements indicating the acidity, basicity, or amphoterism of their oxides. Elements in bold face are acidic, in normal face are basic, in circles are amphoteric in all oxidation states, and in octagons are amphoteric in their lower oxidation states and acidic in their higher oxidation states.

An amphoteric oxide or hydroxide can react with both acids and bases and thus its acidity or basicity depends on whether it is in acid or alkaline solution. We can illustrate this by considering the reactions of aluminium oxide, Al_2O_3, shown in Eqns. 6.8 and 6.9.

$$(basic)\ Al_2O_3(s) + 6H^+(aq) \rightarrow 2Al^{3+}(aq) + 3H_2O(l) \tag{6.8}$$

$$(acidic)\ Al_2O_3(s) + 2OH^-(aq) + 3H_2O \rightarrow 2\left[Al(OH)_4\right]^-(aq) \tag{6.9}$$

In Eqn. 6.8, aluminium oxide acts as a base under acidic conditions with the formation of the element cation, Al^{3+} (this would be present as the aqua-ion $[Al(H_2O)_6]^{3+}$). Under basic conditions as shown in Eqn. 6.9, aluminium oxide behaves as an acid (note that it acts as a hydroxide acceptor rather than a proton donor in this case) with the formation of an element hydroxo-anion.

Having looked at the definitions of acidity, basicity, and amphoterism, it is clear from Fig. 6.3 that elements in the s-block (with the exception of hydrogen) and the lower left of the p-block have basic oxides and hydroxides. Elements in the top right of the p-block are acidic and elements along an approximate diagonal from top left to bottom right tend to be amphoteric. More specifically with regard to amphoterism, the oxides of the elements in circles are amphoteric in all of their oxidation states, whilst those in octagons have acidic oxides in their maximum oxidation states and amphoteric oxides in their lower oxidation states. Group 13 provides a nice example of the trend within a group: B_2O_3 is weakly acidic, Al_2O_3, Ga_2O_3, and In_2O_3 are all amphoteric (although In_2O_3 may be better described as weakly basic), Tl_2O_3 is basic and Tl_2O is strongly basic. The two thallium oxides illustrate the effects of oxidation state which is addressed in more detail below.

There is an obvious parallel between the trend revealed in Fig. 6.3 and that observed for the electronegativity of the elements, and it is on this basis that we can seek an explanation. Thus, it is clear that electronegative elements form acidic oxides, electropositive elements form basic oxides, and amphoterism is observed for elements of intermediate electronegativity. We can understand why this should be the case by considering an individual element hydroxide unit, E–O–H.

If E is very electronegative, the electron density will tend to be polarized towards the E centre, which we can represent as δ^- E–O–H δ^+, or certainly towards the E–O unit, i.e. δ^- (E–O)–H δ^+. This polarization facilitates loss of a proton (H^+), or O–H bond heterolysis, and the negative charge of the resulting oxo-anion is effectively stabilized by the presence of the electronegative element E, i.e. it is not strongly localized at the oxygen atom. In addition, an electronegative element is unlikely to form a cation, E^+. In contrast, if E has a low electronegativity, i.e. if it is an electropositive element, the polarization will tend to be in the opposite sense, i.e. δ^+ E–O(δ^-)–H or, better, δ^+ E–(O–H) δ^-, such that dissociation occurs by heterolysis of the E–O bond to give E^+ and OH^-. All of the above is summarized in Eqns. 6.10 and 6.11. Note that the element E will generally be bonded to other groups as well.

The formation of cations in aqueous solution is a common feature of electropositive, metallic elements.

$$\text{Electronegative E:} \qquad \delta^-(E-O)-H\ \delta^+ \rightarrow EO^- + H^+ \tag{6.10}$$

Electropositive E: $\delta^+ E-(O-H) \; \delta^- \rightarrow E^+ + OH^-$ (6.11)

For elements of intermediate electronegativity, the magnitude of any bond polarization is insufficient to result in the predominance of either acidic or basic behaviour, and amphoteric properties are observed, i.e. acidic or basic behaviour is strongly dependent on pH. However, amphoterism is often seen to be dependent on the element oxidation state. The elements Be, Mg, Al, Ga, and In are amphoteric in all of their oxidation states, though only for the Group 13 elements is there a possibility of more than one oxidation state, and, in fact, stable oxides and hydroxides for the Group 13 elements listed are only found for the +3 state.

With the elements Ge, Sn, Pb, As, Sb, and Bi, the oxides and hydroxides are acidic in their highest oxidation states and amphoteric in lower oxidation states, e.g. arsenic(III) oxide, As_2O_3, is amphoteric whereas arsenic(V) oxide is acidic. We can account for this observation if we recall that element electronegativity is, to some degree, dependent on oxidation state and that the higher the oxidation state, the higher the electronegativity. We noted previously that, although both are basic, thallium(I) oxide, Tl_2O, is more basic than thallium(III) oxide, Tl_2O_3, consistent with the greater electronegativity of Tl(III) compared to Tl(I). Finally, let us note that hydrogen forms an amphoteric oxide, i.e. H_2O.

Another approach to trends in acidity arises from considering the hydrolysis of element chlorides. Thus, if we take a species ECl_x, we can consider its dissociation in water to take place according to Eqn. 6.12 to give chloride anions and aqua-ions of the element cations (where n is typically 6 but will be higher for larger cations and as low as 4 for smaller cations). This is precisely what happens with the Group 1 and 2 chlorides when they dissolve in water, and we would define this as dissociation or dissolution but not as hydrolysis.

$$ECl_x \rightarrow \left[E(H_2O)_n\right]^{x+} + xCl^-$$ (6.12)

As x increases, however, the increasing charge on the expected $[E(H_2O)_n]^{x+}$ cation would result in an increasing tendency towards deprotonation leading to hydroxo species and ultimately, for very electronegative E, oxo-anions as illustrated for a 3+ cation in Eqn. 6.13. This we would classify as hydrolysis.

$$[E(H_2O)_n]^{3+} \rightarrow E(OH)_3 + 3H^+ \text{ and } E(OH)_3 \rightarrow [EO_3]^{3-} + 3H^+$$ (6.13)

We may also consider the trends in the strengths of various acids in aqueous solution as indicated by their pK_a values. Table 6.1 shows the pK_a values for the four oxo-acids of chlorine from which it is clear that the acid strength increases dramatically as the number of oxygen atoms attached to the chlorine atom increases. This is simply a result of more oxygens being able to more effectively delocalize the negative charge (or, in other words, to stabilize and render less basic the conjugate base).

Before leaving this section on oxides and hydroxides we shall draw attention to a number of interesting observations about certain hydroxides and oxo-anions, not particularly related to their acidity but rather to their stability and properties. Thus,

What is illustrated in Eqn. 6.13 is rather generic. Depending on the system, various intermediate species may also be formed and $E(OH)_3$ might be better represented as $[E(H_2O)_{n-3}(OH)_3]$. but that detail is not necessary in terms of the point being made.

Table 6.1 pK_a values for the oxo-acids of chlorine. The formulae are more usually written as HClO, HClO₂, etc. but the form used here better illustrates the structure.

Acid	pK_a
Cl(OH)	7.2
ClO(OH)	2.0
$ClO_2(OH)$	−1
$ClO_3(OH)$	−10

Pauling's first rule for acidity states that for an acid $O_pE(OH)_q$, the pK_a is approximately equal to $8 - 5p$, quite close to the values in Table 6.1 in the examples given. Pauling's second rule states that the successive pK_a values of polyprotic acids (i.e. those with $q > 1$) increase by 5 units for each successive proton loss.

the hypothetical compounds $C(OH)_4$ and $N(OH)_3$ are both unstable and readily dehydrate to give H_2CO_3 and HNO_2 respectively. The reason can be traced to the relative strength of C/N–O multiple bonding as opposed to C/N–O single bonding as described in Chapter 4. A related observation is that the compound $P(OH)_3$ exists not as a P(III) trihydroxide but as the P(V) species $PH(O)(OH)_2$ again as a result of a strong, in this case P–O, multiple bond; this is an example of **prototropic tautomerism** (analogous to keto-enol tautomerism). Another interesting point relates to the properties of the isoelectronic oxo-anions ClO_4^-, SO_4^{2-}, PO_4^{3-}, and SiO_4^{4-} in that there is an increasing tendency as the charge increases towards condensation in the presence of acid (H^+) to give di- or oligonuclear oxo-bridged species, as this relieves the increasing overall charge. Condensed silicates have enormous structural variety and form the basis of a great many minerals. Similar condensation of the oxo-anions of the 2p elements is not as extensive, not being seen at all for C and N, although many condensed borate anions are known; this also reflects a greater stability of C/N–O multiple bonding in the uncondensed species.

A final point on this topic concerns the trend in the acidities of the p-block element hydrides which are shown in Table 6.2.

Table 6.2 pK_a values for some of the p-block element hydrides for Groups 14–17

14		15		16		17	
CH_4	46	NH_3	35	OH_2	16	FH	3
SiH_4	35	PH_3	27	SH_2	7	ClH	−7
GeH_4	25	AsI_3	23	SeH_2	4	BrH	−9
				TeH_2	3	IH	−10

We note that the acidity increases on moving from left to right and also on descending a group. These observations cannot be rationalized by a single principle such as electronegativity or bond strength, but it is clear that for a given row the element–hydrogen electronegativity difference is the dominant factor whereas in a group it is the element–hydrogen bond strength which is the most important. Down a group, weaker bonds result in more ready dissociation, and these values show much the same trend as those measured for the gas phase acidities. Across a row, however, the gas phase acidities are found to be fairly similar, implying that in aqueous solution it is not so much the electronegativity difference which is important but rather the increasing hydration energy associated with smaller, more highly charged anions which have a greater charge-to-size ratio or charge density.

6.3 Lewis acidity of the heavier p-block elements

General points

In Section 6.1 the examples used to illustrate Lewis acidity were BF_3 and SiF_4, and it will be useful to consider the origin of this acidity in both compounds. In the former, the three-coordinate, trivalent boron centre has six electrons (three from itself, one each from the three fluorines), but with four valence orbitals (one s and three p) a total of eight electrons can be accommodated. Thus, boron in BF_3 has a vacant valence orbital which can readily accept an electron pair, through formation of a coordinate or dative covalent bond, resulting in a species with eight valence electrons. We should recognize in BF_3 itself that there will be some interaction between the vacant boron 2p orbital and filled π-type orbitals on the fluorine atoms, i.e. some B–F π-bonding will be present (which can also be considered as an acid–base interaction where the boron is described as a π-acid and fluorine as a π-base). In general, however, the formation of a new σ bond will readily displace any π-bonding electrons, and in BX_3 species which contain X groups not capable of π-donation (e.g. H, alkyl, aryl) this issue does not arise.

In fact, Lewis acidity is a general feature of the chemistry of the Group 13 elements, or indeed any elements with fewer valence electrons than orbitals, at least as far as their covalent chemistry is concerned. We can therefore think of compounds with eight electrons as electronically saturated (or **electron precise**), since all four valence orbitals are associated with two-centre, two-electron bonds; this idea is, of course, the basis of the Lewis octet rule which we shall return to later. Compounds with fewer than eight electrons can therefore be thought of as unsaturated or **electron deficient** with a resulting tendency to form extra bonds to electron donors.

If we now turn to the example of SiF_4, it may not at first be obvious why this molecule should act as a Lewis acid since it is already an eight-electron species and therefore electronically saturated according to the definition given above. Nevertheless, SiF_4 does form an adduct with pyridine (Fig. 6.2) and also with fluoride to give a complex anion of the formula $[SiF_6]^{2-}$ both of which can be considered as twelve-electron species, an apparent violation of the octet rule.

In seeking an explanation for the acidity of SiF_4 we should first consider that the carbon analogue, CF_4, shows no such acidic behaviour and, what is more, it is a general feature of the chemistry of the first-row elements that the octet rule is not violated. Moreover, the acidity of SiF_4 and the apparent violation of the octet rule is not an isolated example but is a common feature of the chemistry of the heavier p-block elements. If we take CF_4 and SiF_4 as examples, two factors which we should initially consider are, firstly, that silicon is bigger than carbon, and secondly, that the Si–F bonds will be more polar than the C–F bonds as the Si/F pair have the greater $\Delta\chi$. As we have seen before, larger atoms can support larger coordination numbers and a more δ^+ silicon centre is certainly likely to be more acidic.

Whilst these ideas of size and polarity are important, they are not a sufficient explanation for the acidity of SiF_4 or, indeed, for the Lewis acidity of a large number of compounds of the heavier p-block elements. This is because the bonding between silicon and the pyridine ligands in $[SiF_4(py)_2]$ is not ionic or electrostatic; there is a high degree of covalency so there must be some suitable orbitals associated with the silicon centre. In other words, SiF_4 must have some low-energy or low-lying vacant orbitals with which it can form additional covalent interactions

With regard to π-bonding in the boron trihalides, it is generally supposed that the smaller the halide, the greater should be the B–X π-bonding (due to a more favourable orbital size match) which is the reason offered to account for the increase in the Lewis acidity of boron trihalides in the order $BF_3 < BCl_3 < BBr_3 < BI_3$, the opposite trend to that expected on the basis of element electronegativity differences.

The borate anion BO_3^{3-}, which is isoelectronic with BF_3, is also a Lewis acid, whereas the isoelectronic carbonate and nitrate anions, CO_3^{2-} and NO_3^-, are not. This reflects the lower difference in electronegativity between C/N and O, and also the stronger C–O and N–O π-bonding.

Size and bond polarity can also account for the related fact that SiF_4 and $SiCl_4$ are readily hydrolysed whereas CF_4 and CCl_4 are not. Thus, H_2O can coordinate to the silicon centre whereas this is not possible in the case of carbon. Subsequent HF or HCl elimination is facilitated by the close proximity of the H and F or Cl atoms in the silicon coordination complex and also by the more δ^- F or Cl centres.

There is an interesting aspect of Lewis acidity in relation to the stability of the group maximum oxidation state for the 4p elements (Section 4.2). For example, $AsCl_5$ decomposes above −50°C into $AsCl_3$ and Cl_2, but the complex $[AsCl_5(OPMe_3)]$ is stable up to +50°C which suggests that the normally facile decomposition modes are kinetic in origin and can be arrested by effectively saturating the coordination sphere of the central element.

with Lewis bases. With regard to the nature of these orbitals, we shall consider two particular models.

Vacant d orbitals

A traditional view of bonding in a molecule like $[SiF_4(py)_2]$ is to assume that all bonds are of the standard two-centre, two-electron type and therefore that there are twelve electrons associated with the silicon centre (four Si–F bonds and two dative Si–N bonds). This feature was mentioned previously and is why the compounds are said to violate the octet rule. Since silicon only has four valence orbitals, which can accommodate a maximum of eight electrons, other orbitals must be involved, and it is assumed that these are vacant, low-lying 3d orbitals. It is then possible to use two 3d orbitals (usually d_{z^2} and $d_{x^2-y^2}$) as well as the 3s and three 3p orbitals to construct a set of six d^2sp^3 hybrids with which to form six two-centre, two-electron bonds. Moreover, the hybrids chosen are equivalent and point to the vertices of an octahedron which accounts for both the shape and the six equal Si–F bond lengths in $[SiF_6]^{2-}$. In employing vacant d orbitals, the problems with violation of the octet rule (hypervalence, see Section 4.1) in compounds of the heavier p-block elements compounds are circumvented, and the lack of any such compounds in the chemistry of the 2p elements is readily understood in view of the absence of 2d orbitals.

This model employing d orbitals is quite useful and has achieved considerable currency amongst inorganic chemists. Furthermore, we can develop the ideas to account for the different bond lengths found in a trigonal bipyramidal molecule such as PF_5, an example of a hypervalent, 10-electron phosphorus compound. In PF_5, two-centre, two-electron bonding requires the use of one d orbital to construct a set of five dsp^3 hybrids, although it is not possible to generate five equivalent hybrids. In fact, the hybrids are constructed in two sets; an sp^2 set which form bonds to the three equatorial fluorines, and a dp set which are involved in bonding to the axial fluorine atoms. Since electrons are most tightly held in s orbitals (since they are lower in energy), and hybrid orbitals with an s component tend to be smaller (less diffuse) than those without, especially for orbitals where $n > 2$, the consequence is that bonds to orbitals which contain any degree of s character tend to be shorter. This model accounts in a simple way for the observed shorter equatorial P–F bonds in PF_5, though it should be noted that significantly shorter equatorial bonds tend to be a feature of compounds with smaller central element centres; in trigonal bipyramidal $BiMe_5$, for example, all Bi–C bonds are equal in length.

A final molecule which we shall consider in this section is the triiodide anion, I_3^-. This ion is readily formed from diiodine, I_2, and iodide, I^-, which can be viewed as an acid–base reaction where I_2 is the Lewis acid. We can assume that I_2 is a Lewis acid due to the presence of vacant d orbitals, and further, that the hybridization of the resulting central iodine in the linear I_3^- is dsp^3 with bonds formed to the two terminal iodines using the axial dp set. The three lone pairs on the central iodine then reside in the equatorial sp^2 orbitals. The longer observed I–I bonds

The deconstruction of dsp^3 hybrids into an sp^2 set and a dp set also accounts for the axial site preference in trigonal bipyramidal molecules; electronegative atoms or groups will tend to bond to hybrids in which the electrons are less tightly held, i.e. those with least or no s-character. This is an example of Bent's rule, which is explained later. Hybrid orbitals are discussed in more detail in Winter (2016).

This is a good place to note that the compound of the formula TlI_3 is actually a compound of Tl(I) rather than Tl(III), i.e. it comprises Tl^+ and I_3^- ions rather than Tl^{3+} and three I^- ions; this is another example of the inert pair effect in thallium chemistry.

in I_3^- compared to that in I_2 can also be accounted for, since in the former they involve dp hybrids, whereas in the latter the I–I bond is formed from sp hybrids (or at least hybrids with some s character).

In considering I_3^-, we described the structure as linear; this is what is observed, and it is predicted on the basis of valence shell electron pair repulsion (VSEPR) theory which has been extensively developed by Gillespie and which we shall look at in more detail in Chapter 7. Indeed, VSEPR is an extremely powerful model for rationalizing the structures of molecules in the p-block. It explicitly employs the concept of two-centre, two-electron bonds, or more generally, electron-pair domains (lone pairs are considered as well), and thus, implicitly, the concept of d orbital hybridization in hypervalent compounds. Herein, however, lies one of the problems with the d orbital model for Lewis acidity. The structures of hypervalent molecules can be accounted for on the basis of VSEPR and a certain hybridization can then be assumed, i.e. dsp^3, d^2sp^3, etc., but we cannot easily use the d orbital model of itself to predict structure. The central iodine in I_3^- can be described as dsp^3 hybridized, but the use of one particular d orbital and the resulting deconstruction to sp^2 and dp sets is rather arbitrary. It is the same as assuming (wrongly) that methane is tetrahedral *because* it is sp^3 hybridized; in fact, methane is tetrahedral because this is the lowest energy configuration for the molecule, and sp^3 hybridization can then be used (as one model) to describe the bonding.

In other words, we cannot use d orbitals in themselves to account for the structures of molecules such as I_3^-, but rather must use them in conjunction with VSEPR. Whilst this is not a serious problem, a second problem makes the d orbital argument much less tenable since it is now generally accepted that d orbitals are too high in energy to be used effectively in bonding in the p-block. An alternative explanation for the Lewis acidity of the heavier p-block elements is therefore desirable, and as we shall see in the following section there is a model which accounts for Lewis acidity, structure, and bond lengths.

It should be stressed that hybridization need only be introduced after the fact in discussions of VSEPR, it is not really required. However, if we wish to equate electron-pair domains with particular hybrids, d orbitals are needed for compounds with more than four electron pairs.

Theoretical studies confirm that vacant d orbitals are too high in energy to play an important role in bonding in the sense discussed in this section. If d orbitals are not involved, however, why do we not see an extensive hypervalent chemistry of the 2p elements? The reason is that the small size and high electronegativity of the 2p elements effectively prevents large coordination numbers and high formal charges.

Vacant σ^* orbitals

Let us consider I_3^- again. We can, in fact, construct a bonding or molecular orbital picture for this molecule in which d orbitals are unnecessary. If we assume that the central iodine has three lone pairs in three sp^2 orbitals, there is one remaining p orbital which can be used to form a multicentre bond with suitably hybridized σ orbitals of the two terminal iodines as represented in Fig. 6.4. A three-

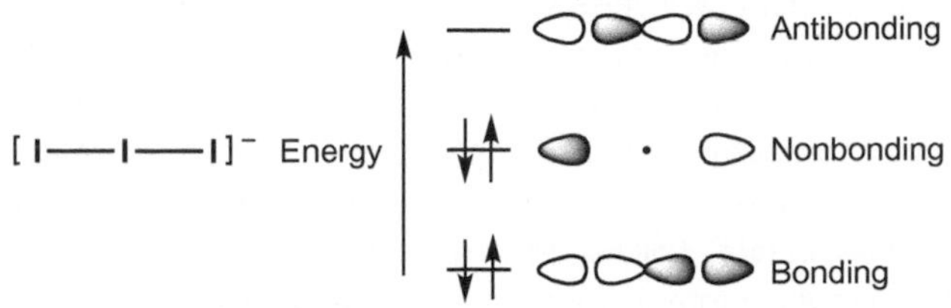

Fig. 6.4 An orbital interaction diagram for I_3^-.

A much more in-depth discussion of many aspects of molecular orbital theory is given by Winter (2016).

This multicentre approach to bonding was first discussed as early as the 1950s.

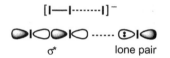

Fig. 6.5 The I–I σ*-orbital derived from the overlap of two p_z orbitals.

Fig. 6.6 Orbital interactions between I_2 and an approaching I^-.

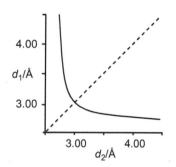

Fig. 6.7 A graph of the correlation of the I–I and I–I bond distances in I_3^- anions; the graph is symmetrical about the dashed diagonal line.

It should be noted that the curve in Fig. 6.7 can also be considered as a reaction coordinate for the nucleophilic displacement of iodide from diiodine by iodide, i.e. $I–I + I^- \rightarrow I^- + I–I$. More detail on these primary *vs* secondary bond length correlations (and **secondary bonding** in general) and models of reaction coordinates is provided in Alcock (1990).

atom, three-orbital interaction results in which the bonding and non-bonding orbitals are filled and for which the net I–I bond order is one half. The reduced bond order of this three-centre, four-electron bond thus neatly accounts for the longer I–I bonds without recourse to d orbitals. Moreover, the electron count around the central iodine is effectively eight, since there are only three lone pairs and one bonding pair associated with the central iodine. The pair of electrons in the non-bonding orbital are localized on the terminal iodines and do not contribute to the electron count at the central iodine, so it can be seen that the octet rule is not really violated at all. An entirely equivalent description can be used to describe the bonding in the isoelectronic XeF_2. Moreover, this sort of bonding argument ignoring d orbitals is quite general and can be used, for example, to account for the longer axial P–F bonds in PF_5. Thus, the equatorial P–F bonds are two-centre, two-electron bonds formed by sp^2 hybrids on the phosphorus whilst the two axial P–F bonds are described in exactly the same way as the I–I–I bonding in I_3^-. The equatorial P–F bonds therefore have a bond order of 1 whereas the axial P–F bonds have a bond order of ½. On this basis, the effective electron count around the phosphorus atom is also eight.

Thus, we can abandon d orbitals in describing the bonding in I_3^-, but we must still account for the Lewis acidity of I_2. In other words, we must consider the nature of the low-lying vacant orbital(s) of I_2. The most appropriate is the I–I σ* orbital (Fig. 6.5) which is expected to be quite low in energy due to the large and diffuse nature of the iodine orbitals (the corollary is that the bonding orbital will be high in energy, which accounts for the relative weakness of the I–I bond).

If we consider the interaction between I_2 and I^-, it is clear that maximum overlap will occur when the iodide approaches *trans* to the I–I bond (Fig. 6.6). Moreover, as the I–I to I^- distance decreases, the population of the I–I σ* orbital will increase with a concomitant lengthening of the I–I bond. In the extreme where both I–I distances are the same, the delocalized bonding scheme described in Fig. 6.4 is appropriate.

Thus, not only does the σ* model account for the observed geometry of I_3^- and for the longer I–I bonds observed in I_3^- *vs* I_2, it also suggests that a correlation should exist between the I to I distances in the unsymmetrical species, I–I····I, i.e. as the secondary I····I bond, as it is often called, gets shorter, the primary bond gets longer. This is exactly what is observed in a range of structures involving the I_3^- anion; the I–I and I····I distances are correlated and are found to lie on the curve shown in Fig. 6.7.

Thus, multicentre bonding and the σ* orbital model of Lewis acidity can account for all of the observations traditionally rationalized on the basis of d orbital participation. Moreover, details of the structures of Lewis acid–base complexes of the p-block elements are understood in a more straightforward manner than they are based on models using d orbitals. This is especially well illustrated in the case of the solid-state structures of the tri-iodides of arsenic, antimony, and bismuth. All three compounds are isomorphous and can be considered as a close-packed array of iodines with As, Sb, or Bi in the octahedral holes (1/3 of the octahedral holes, in fact). For AsI_3 there are three short, mutually *cis* As–I bonds with three longer As····I interactions *trans* to these bonds. In SbI_3 the basic structure is the same except that the ratio of the primary to the secondary bonds is smaller; for BiI_3 the ratio is 1 and all Bi–I bonds are the same length (Fig. 6.8). Note that the *trans* disposition of the secondary bonds to the primary bonds is exactly what would be predicted on the basis of the σ* orbital model.

Moreover, the increasing Lewis acidity of BiI_3 *vs* AsI_3, as exemplified by the more symmetrical structure, i.e. a smaller ratio of primary to secondary bond lengths, is consistent with a lower-energy σ* orbital formed by overlap between two large atoms with large diffuse orbitals. We should also note that this model would lead us to expect a regular octahedral structure for the BiI_6 unit, whereas the VSEPR predicted structure would be ambiguous due to the presence of a lone pair of electrons (we shall return to this in Chapter 7); each bismuth centre has seven electron pairs, and the stereochemical activity or not of the lone pair is always problematical for these types of structure.

We should note that the trend in the structures of AsI_3, SbI_3, and BiI_3 is essentially from covalent to polymeric, but this is slightly counter to what we would expect on the basis of electronegativity differences (Table 2.5). However, in no case is the element electronegativity difference large, and it is therefore not such a good guide in accounting for this trend. What is more important here is the trend to increasing Lewis acidity as the group is descended and hence the increasing trend toward strong secondary bonding interactions and polymeric structures. We should note also that the electronegativities of Bi (2.02) and I (2.66) are such as to place the compound near the metalloid region, and consistent with this is the fact that BiI_3 is a semiconductor.

We shall conclude this section by noting that employing σ* orbitals readily accounts for the Lewis acidity of the compounds of the heavier main group elements, since it is for these elements that orbitals are large and diffuse and for which σ* orbitals would be expected to be low in energy. For elements of the 2p row, Lewis acidity is not observed because orbital overlap is good and the σ* orbitals are therefore much higher in energy. Moreover, heavy-element iodides would be expected to exhibit the most Lewis acidic behaviour (as in the iodides of As, Sb and Bi), since two large elements are involved and bonding occurs between two large and diffuse orbitals (but see below).

Ideas relating to hypervalence and the octet rule are still a matter of much debate, which is why the two models are covered here in some detail. Thus, the neglect of d orbitals in molecular orbital terms does not tend to lead to violations of the octet rule as we have seen for I_3^- and PF_5 (and a perfectly adequate MO description of the bonding in octahedral SF_6 can be derived without the need for d orbitals, as we will show in Chapter 8), and as such the term 'hypervalence' or 'hypervalent' becomes redundant. However, if we do consider the bonding in, for example, PF_5 or SF_6 to comprise two-centre, two-electron P–F or S–F bonds, a total of ten and twelve valence electrons respectively are required which clearly violates the octet rule, and such compounds are still often termed 'hypervalent', as noted in Section 4.1. Nevertheless, the usefulness of the term 'hypervalence' has been questioned by Gillespie (although for different reasons) who would rather it be abandoned. In its place, Gillespie would emphasize the octet rule and a duodecet rule for second-row molecules like CF_4, and for third- and higher-row molecules like SF_6 respectively, both of which are chemically stable and have many properties in

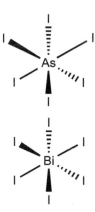

Fig. 6.8 The coordination geometry around arsenic and bismuth in the solid-state structures of AsI_3 and BiI_3 respectively.

The structure of BiI_3, with symmetrical six-fold coordination around the bismuth centres, is the same as that of $AlCl_3$.

For the trihalides of As, Sb, and Bi (EX_3) it is generally the case that for a E–X–E bridging situation, for a given E the bridge is more symmetric as X becomes less electronegative, and that for a given X the bridge is more symmetric as E becomes less electronegative. In both cases there is a trend to increasing metallic or metalloid behaviour depending on the absolute electronegativities of E and X.

The low energy of the σ* orbitals is also important in understanding the observation that iodides are often coloured whereas chlorides and fluorides are not, as noted in Section 5.5. The n→σ*-transitions which give rise to the colour tend to occur in the visible region for iodides whereas they are in the ultraviolet for the lighter halides, and this also accounts for the photosensitivity of many heavy-metal iodides.

common. Exceptions would then be the Lewis acidic species like the six-electron BF_3 and the ten-electron PF_5, both of which readily accept fluoride to form the octet and duodecet species $[BF_4]^-$ and $[PF_6]^-$. Quite where this would leave SiF_4, however, is unclear; it has an octet of electrons like CF_4 but can bond to two extra fluorides to give the duodecet compound $[SiF_6]^-$. Also, there are many compounds which formally have fourteen valence electrons, as we shall see in Chapter 7.

As a final example of the application of σ* orbitals, we will briefly note how they can be employed in understanding the bonding in species such as Cl_3PO or phosphine oxides generally and related compounds such as ylides. We looked at Cl_3PO in Chapter 4, and will consider it again in Chapter 7 when looking at how VSEPR deals with molecules that can be represented with double bonds of this type. Earlier ideas and many older texts will have considered the bonding in such species to involve π-donation from oxygen lone pairs into vacant phosphorus 3d orbitals, as shown in Fig. 6.9(a). A better description, however, involves π-donation into the P–Cl σ* orbitals, as shown for one particular interaction of this type in Fig. 6.9(b). Although we shall not dwell on the matter here, π-donation into P–X σ* orbitals in X_3PY species generally can also account for changes observed in the P–X bond distances and in the X–P–X angles as X and Y are varied; if Y is a transition-metal fragment, this also offers a good model for how phosphines and related ligands bond to transition metals and act as π-acceptors.

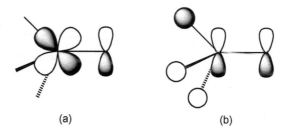

(a)   (b)

Fig. 6.9 For X_3PO, (a) π-donation from O into a phosphorus d orbital or (b) π-donation from O into a P–X σ*-orbital.

6.4 Hard and soft acids and bases

In concluding this chapter we should say something about the concept of **hard and soft acids and bases** which owes much to the work of R. G. Pearson. In general, both acids and bases can be classified as either hard or soft, and it is found that hard acids preferentially bond to (or form adducts with) hard bases and soft acids bond to (or form adducts with) soft bases. A selection of hard and soft acids and bases is shown in Table 6.3.

A simple and useful explanation for the attraction of hard to hard and soft to soft is as follows. Hard acids (electron-pair acceptors) have no low-energy or low-lying unfilled orbitals (i.e. the LUMO is high in energy) and therefore do not readily form covalent bonds with bases. Hard bases (electron-pair donors) have low-energy filled orbitals (i.e. the HOMO is low in energy) and so do not readily

HOMO means highest occupied molecular orbital. and LUMO means lowest unoccupied molecular orbital.

Table 6.3 A selection of hard and soft acids and bases together with some borderline cases for s- and p-block species

Acids		
hard	**borderline**	**soft**
H^+, Li^+, Be^{2+}, Mg^{2+}, Al^{3+}, Ga^{3+}, In^{3+}, As^{3+}, BF_3, $AlMe_3$, $AlCl_3$, SO_3	Sn^{2+}, Pb^{2+}, Bi^{3+}, BMe_3, SO_2, NO^+, GaH_3	Tl^+, Hg^{2+}, Tl^{3+}, $TlMe_3$, BH_3, $GaMe_3$, $GaCl_3$, $InCl_3$, I^+, I_2

Bases		
hard	**borderline**	**soft**
H_2O, OH^-, F^-, SO_4^{2-}, Cl^-, CO_3^{2-}, ClO_4^-, NO_3^-, NH_3	N_3^-, Br^-, NO_2^-, SO_3^{2-}, N_2	H_2S, I^-, SCN^-, R_3P, CN^-, CO, H^-

form covalent bonds with acids, but since the acids and bases are usually positively and negatively charged respectively, they tend to bond strongly to each other through ionic or electrostatic interactions.

In contrast, soft acids do tend to have low-energy or low-lying unfilled orbitals (a low-energy LUMO) which will readily enter into covalent bond formation, and soft bases tend to have high-energy or high-lying filled orbitals (a high energy HOMO) which will also easily form covalent bonds. The combination of soft acids with soft bases, therefore, is the one which maximizes a covalent interaction. These are extremes, of course, and those classified in Table 6.3 as borderline are intermediate in their acid/base properties which can be compared with amphoterism.

> Remember that covalent interactions are strongest when the energies of the contributing orbitals are similar.

The general principle of hardness (usually given the symbol η) can be quantified (for atoms, ions, or molecules) as shown in Eqn. 6.14, where I is the first ionization energy, and A is the first electron affinity.

$$\eta = (I-A)/2 \tag{6.14}$$

In orbital terms, I and A correspond to the energies of the HOMO and LUMO respectively (assuming Koopmans' theorem holds), such that η can also be expressed in terms of the energies of the HOMO and the LUMO, as shown in Eqn. 6.15.

$$\eta = (E_{LUMO} - E_{HOMO})/2 \tag{6.15}$$

As we saw in Chapter 2, orbital energies are not the same as ionization energies or electron affinities for polyelectronic systems (i.e. Koopmans' theorem does not hold), though the differences are often small. Eqn. 6.15 does, however, serve to highlight the fact that increasing hardness corresponds to an increasing energy separation between the HOMO and the LUMO. There is a so-called **principle of maximum hardness** which states that chemical systems will tend to be as hard as possible and that chemical reactions will occur in a manner which maximizes the hardness of the products, i.e. affords products with the lowest-energy HOMOs.

The opposite of hardness, i.e. softness, is given by the expression $\sigma = 1/\eta$. A similar property, the chemical potential, μ, defined as in Eqn. 6.16, is also encountered in this context; the negative of μ is equal to the Mulliken electronegativity (χ_M) mentioned in Chapter 2. We can therefore think of species with a high chemical potential as being those with high-energy HOMOs, i.e. those which are electropositive.

$$-\mu = (I+A)/2 = \chi_M \qquad (6.16)$$

Exercises

1. Predict the structure of and describe the bonding in the product formed from the reaction between diborane, B_2H_6, and ammonia, NH_3.

2. Write down the reactions expected between K_2O and water and between Cl_2O and water, and comment on the acidity or basicity of the resulting solutions.

3. Consider whether SF_4 violates the octet rule.

7 Structure

7.1 Electron counting rules

Before looking at some selected periodic trends in relation to structure, we shall look briefly at the **octet rule** and touch on some other electron counting rules generally.

The octet rule was first formulated by Gilbert Lewis and Irving Langmuir in the early twentieth century, albeit building on the earlier work of others, and we have mentioned it several times throughout this text. Octet refers to 8, of course, and this is an important number especially for the s- and p-block elements. Briefly stated, the importance of the number 8 arises since, for these elements, we have one valence s and three valence p orbitals which, when involved in covalent bonding, will result in four bonding orbitals and four antibonding orbitals (or a reduced number corresponding to the number of lone pairs or non-bonding orbitals); 8 electrons will therefore fill the bonding/non-bonding orbitals completely. In ionic structures, it is about losing or gaining enough electrons to achieve the filled shell of 8; think about Na losing one electron to form Na^+ and F gaining one electron to form F^-. Ions which achieve an octet by losing or gaining electrons or atoms which achieve an octet by sharing electrons covalently are said to be **electron precise** and to obey the octet rule.

Covalent compounds which have fewer valence electrons than valence orbitals are described as being **electron deficient**, and we have encountered some examples in previous chapters. These compounds do not, to a first approximation, obey the octet rule, and we have seen that their structures and chemistry are dominated by finding more electrons so that they do obey (or tend towards obeying) the rule; think of bridging hydrogens in B_2H_6, B–X π-donation in BX_3 trihalides and adduct formation in $Me_3N \rightarrow BF_3$.

What about compounds which seemingly have more than 8 electrons, many of which we have met and discussed in previous chapters? In terms of the discussion above, were the number of electrons to exceed 8, these would have, in covalent compounds, to go into antibonding orbitals, and it is this which gives us the insight into why this rule works, and, why indeed, electron counting rules generally, work (and why, in some cases, they do not work). As we mentioned in Chapter 4 when discussing oxidation states, orbital overlap tends to be good in the p-block which results in low-energy bonding orbitals and high-energy antibonding orbitals or, in other words, a large HOMO-LUMO gap. In such circumstances, it is therefore favourable to fill the bonding (and any non-bonding) orbitals but not to populate the antibonding orbitals, and this is the basis for the rule.

In fact, a large HOMO-LUMO gap is the basis for all electron counting rules. Other examples are the Hückel 4n+2 rule for aromatic compounds, the 18-electron rule in transition metal chemistry, and Wade's rules (later developed into Wade–Mingos rules or Polyhedral Skeletal Electron Pair Theory, PSEPT) for polyhedral boranes and transition metal carbonyl clusters. All of these rules work (or are obeyed) because of large separations in energy between filled and unfilled orbitals.

For those compounds which do seemingly violate the octet rule, the so-called hypervalent compounds which apparently have 10, 12, or even 14 electrons, we discussed two bonding models in Chapter 6. Either more orbitals are involved, these being vacant d-orbitals, or we look at the electronic structure in terms of multicentre bonding. In the former case, we still have filled bonding and unfilled antibonding orbitals, and Gillespie would go as far as to propose a duodecet rule for 12-electron compounds. However, for the ten- and twelve-electron compounds we considered in Chapter 6, it was argued on the basis of a multicentre bonding model that with non-bonding orbitals which have a node at the central atom, the total number of electrons at the central atom never actually exceeds 8 in the first place, invoking d orbitals is unnecessary, and problems with any violation of the octet rule largely disappear.

7.2 Valence shell electron pair repulsion theory (VSEPR)

The molecules formed by elements of the p-block often appear to present a bewildering array of structures. For example, if we consider some of the trifluorides, BF_3 is trigonal planar, NF_3 is trigonal pyramidal, and ClF_3 is T-shaped. Moreover, PF_3 is found to be 'more' pyramidal than NF_3, i.e. the F–P–F angles are smaller than the F–N–F angles. A successful theory of molecular structure must account for all of these observations in a simple and elegant manner, and a theory which certainly meets these two criteria is known as *valence shell electron pair repulsion* (*VSEPR*) *theory* or sometimes by the names of its two most recent proponents and developers, the Gillespie–Nyholm rules.

VSEPR is covered in many basic inorganic chemistry texts and, in most detail and with many examples, in Gillespie and Hargittai (1991). There are some important trends in molecular structure which we shall want to address, however, and for that purpose it will be useful to present the basic rules of VSEPR and to note the reasons behind the common shapes which molecules adopt. Some of the exceptions to what is predicted by VSEPR rules can also be illuminating, and we shall consider a couple of examples.

VSEPR, as the name implies, is a set of rules with which to predict or rationalize the structures of molecules based on the repulsion between pairs of valence electrons (both bonding pairs and non-bonding pairs); core electrons are completely ignored since their distribution around the nucleus is spherical and

In transition metal chemistry, the hexa-aquo ions $[M(H_2O)_6]^{2+}$ have electron counts ranging from 14 to 22 for Ti to Zn, but metal carbonyls generally obey the 18-electron rule. The reason is that H_2O is a weak field ligand whereas CO is a strong field ligand resulting in small and large splittings between t_{2g} and e_g orbitals respectively.

highly unpolarizable. We should bear in mind that the concept of two-centre, two-electron bonds is used explicitly in this regard, and hybridization schemes can subsequently be employed once a structure is established, but we should recognize that VSEPR is a model for predicting the arrangement of atoms, not for describing details of electronic structure.

The geometries adopted for two to six pairs of electrons are linear, trigonal-planar, tetrahedral, trigonal-bipyramidal, and octahedral respectively: illustrative examples where all of the electron pairs are bonding pairs are shown in Fig. 7.1.

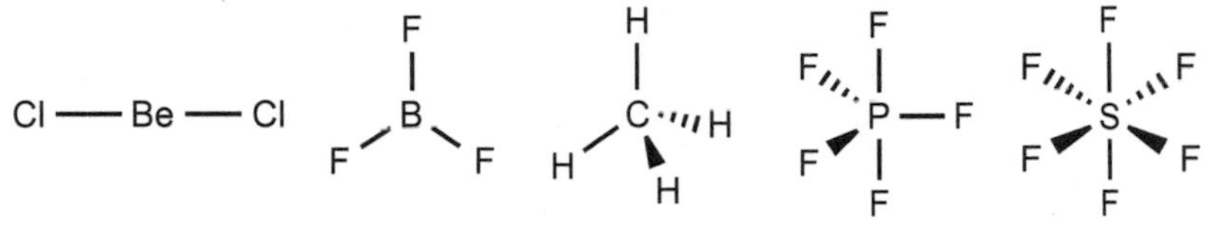

VSEPR is only really appropriate for s- and p-block molecules, since for compounds of the d-block elements the assumption of unpolarizable, spherical cores is not justified. Where a spherical core is appropriate, as is the case for d^0, high-spin d^5, and d^{10} electron counts, VSEPR can be used.

Recent discussions of VSEPR focus more on electron pair domains (bond pair, lone pair, etc.) and the space they occupy rather than on electron pair repulsions. See, for example, Gillespie *et al. J. Chem. Educ.*, 1996, **73**, 622, and *Angew. Chem.*, 1996, **35**, 495.

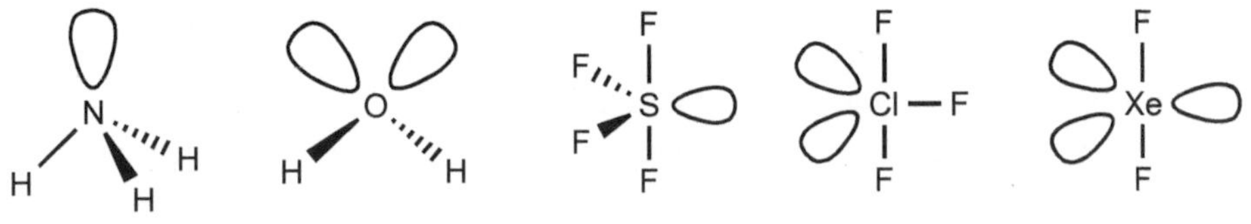

Fig. 7.1 Examples of molecules with two to six bonding electron pairs.

Non-bonding electron pairs also occupy a distinct region of space and influence the structure accordingly. Thus, NH_3 is pyramidal with three bonding pairs and one lone pair and H_2O is bent or V-shaped with two bond pairs and two lone pairs (Figs. 7.2(a) and (b)). We should also note the general rule that non-bonding pairs occupy more space around a central atom than a bonding pair since they are associated only with that atom and are not 'stretched out' between two atoms as is the case with a bonding pair. Thus, whereas in CH_4 the angles subtended at the central atom are 109.5°, i.e. those of the regular tetrahedron, the H–N–H angles in NH_3 are 107.5° and the H–O–H angle in H_2O is 104.5° which together demonstrate the trend towards decreasing interbond angles as the number of lone pairs is increased. In terms of repulsion between electron pairs: lone pair–lone pair > lone pair-bond pair > bond pair–bond pair.

For a trigonal bipyramidal geometry, lone pairs always go into the equatorial sites as this minimizes the 90° contacts (each axial site has three 90° contacts to the equatorial positions whereas each equatorial site has only two 90° contacts to the axial positions; contacts greater than 90° are considered relatively unimportant). This is illustrated with the structures of SF_4 (equatorially vacant trigonal bipyramidal or disphenoidal), ClF_3 (T-shaped), and XeF_2 (linear) (Figs. 7.2(c)-(e)) for one, two, and three lone pairs respectively.

For an octahedral geometry derived from six electron pairs, a molecule with five bonding pairs and one lone pair adopts a square-based pyramidal geometry,

Fig. 7.2 The structures of (a) NH_3, (b) H_2O, (c) SF_4, (d) ClF_3, and (e) XeF_2 showing the lone pairs.

whereas for four bond pairs and two lone pairs, the latter occupy *trans* sites and a molecule like XeF_4 is square planar. These rules and the shapes are summarized in Table 7.1 where EX_nZ_m represents a central atom, E, surrounded by nX groups and m lone pairs, Z (Z does not mean the same here as it does in the CBC model discussed in Chapter 4).

In some solid-state structures, EX_4Z centres adopt a square-based pyramidal geometry rather than the disphenoidal geometry seen for SF_4; SnO (see Fig. 5.14) is an example in which the oxygens occupy the basal sites and the lone pair resides at the apex of the pyramid. There are also a few cases of EX_5 centres which are square pyramidal rather than trigonal bipyramidal; the anions $[InCl_5]^{2-}$ and $[TlCl_5]^{2-}$ are examples.

Table 7.1   Geometrical arrangements for EX_nZ_m molecules

No. of electron pairs	Electron pair arrangement	n	m	Type	Molecular shape	Examples
2	Linear	2	0	EX_2	Linear	$BeCl_2$
3	Trigonal planar	3	0	EX_3	Trigonal Planar	BF_3
		2	1	EX_2Z	V-shaped	$SnCl_2$
4	Tetrahedral	4	0	EX_4	Tetrahedral	CH_4
		3	1	EX_3Z	Trigonal Pyramidal	NH_3
		2	2	EX_2Z_2	V-shaped	H_2O
5	Trigonal bipyramidal	5	0	EX_5	Trigonal Bipyramidal	PF_5
		4	1	EX_4Z	Disphenoidal	SF_4
		3	2	EX_3Z_2	T-shaped	ClF_3
		2	3	EX_2Z_3	Linear	XeF_2
6	Octahedral	6	0	EX_6	Octahedral	SF_6
		5	1	EX_5Z	Square pyramidal	IF_5
		4	2	EX_4Z_2	Square planar	XeF_4

Molecules with double or triple bonds require an additional rule which states that a multiple bond has the same gross stereochemical influence as a single bond. For example, in $POCl_3$ there are five bonding pairs, three P–Cl pairs, and two pairs associated with the P=O double bond (based on the representation shown in Fig. 4.2(c)). Since the latter two pairs act as one, i.e. they are shared between the same two atoms, the geometry of $POCl_3$ is determined by four pairs rather than five, and the molecule is tetrahedral. In general, multiple bond pairs exert a greater repulsion at the central atom than lone pairs or single bond pairs (because they have two or more pairs of electrons), and the interbond angles deviate from idealized values accordingly. Also, and unsurprisingly, in trigonal bipyramidal arrangements, multiply bonded atoms occupy one or more of the equatorial sites for the same reason as do lone pairs. A list of general structural types involving double bonds up to and including a total of six electron pairs is presented in Table 7.2.

An alternative way of dealing with multiple bonds is as follows. To take the example of $POCl_3$ again, the P=O double bond could be drawn as $P^+–O^-$ (see Fig. 4.2(b)). The phosphorus now has one electron less, and the O^- only

Table 7.2 Structural types for molecules involving double bonds

Total no. of electron pairs	No. of double bonds	No. of lone pairs	Shape	Example
4	1	1	V-shaped	NOCl
	1	0	Trigonal planar	$COCl_2$
	2	0	Linear	CO_2
5	2	0	Trigonal planar	FNO_2
	2	1	V-shaped	SO_2
	1	0	Tetrahedral	$POCl_3$
	1	1	Trigonal pyramidal	$SOCl_2$
	1	2	V-shaped	FClO
6	3	0	Trigonal planar	SO_3
	2	0	Tetrahedral	SO_2Cl_2
	2	1	Trigonal pyramidal	$FClO_2$
	2	2	V-shaped	XeO_2
	1	0	Trigonal bipyramidal	SOF_4
	1	1	Disphenoidal	IOF_3
	1	2	T-shaped	$XeOF_2$

contributes one electron (O^- only needs to share one electron to get an octet) so the total number around the phosphorus is now eight electrons or four pairs which predicts the same tetrahedral structure. Accounting for which pair takes up more space and the corresponding effect on bond angles might be a little different according to the approach taken, but as we shall see later, we have to be careful in using VSEPR to account for bond angles.

These two complementary approaches to dealing with elements such as O show that the VSEPR model only really needs to be based on the number of atoms bonded to the central element plus the number of lone pairs. The total number of electron pairs differs according to the model used.

Another aspect of VSEPR which we should consider can be illustrated by examining the structure of PF_2Cl_3. This molecule is a simple derivative of PF_5 and has the expected trigonal bipyramidal structure, but the question remains as to how the atoms are arranged since a number of isomers is possible depending on which atoms are in the equatorial sites and which are axial. The structure observed is the one with both fluorine atoms in axial sites, and this observation turns out be quite general in that electronegative atoms or substituents prefer axial positions in a trigonal bipyramid. The origin of this site preference can be traced to the smaller effective size of electronegative atoms at the central element due to the resultant polarization of the bonding electron density away

Consider also the two approaches for sulfate, SO_4^{2-}: (i) with four S=O bonds; 6 electrons from S, 8 from four oxygens and the 2– charge gives 16 electrons or 8 pairs, but four double bonds means the structure is determined by four pairs not eight; (ii) with four S^+–O^- bonds; 2 electrons from S (it is formally now 4+), 4 from four O^- atoms and the 2– charge gives 8 electrons or 4 pairs: both predict a tetrahedral structure.

from the central atom. This enables electronegative atoms to more readily tolerate the axial site with its three 90° contacts (the opposite of why a lone pair prefers an equatorial site).

An alternative explanation is embodied in Bent's rule which concentrates more on the electronic origins of this site preference. Recall that in Section 6.3, the hybrid orbitals appropriate for a trigonal bipyramid were discussed in terms of a dsp^3 set which could be broken down into equatorial sp^2 and axial dp subsets. These subsets are not equivalent, and electrons will have different energies depending on which subset they are in. Specifically, electrons will be more tightly bound in the sp^2 subset rather than the dp set since orbital energies are such that s is lower than p which is lower than d, as we saw in Chapter 1. Alternatively, we can assign an effective electronegativity to an orbital or hybrid, and it is clear that in the present example, the sp^2 set can be considered as more electronegative than the dp set since the electrons are more tightly held. The result is that an electronegative atom such as fluorine tends to bond to the least electronegative hybrid (given a choice), which in this case is dp since there is less 'competition' for electrons thereby accounting for the observed axial site preference. A third explanation can be offered based on the molecular orbital diagram shown in Fig. 6.4 in Chapter 6. Here the linear three-atom arrangement comprising the two axial atoms and the central atom have a non-bonding orbital localized solely on the axial substituents. This resultant build-up of electron density on the axial atoms (it is not shared with the central atom) is therefore best accommodated when the axial atoms have a large electronegativity.

So far, we have mostly mentioned compounds with up to six electron pairs. There are some with seven, eight, and even nine pairs which we should at least note, not least because with one particular system, we meet our first exception to VSEPR rules.

Seven bonding pairs is exemplified by IF_7 which has a pentagonal bipyramidal structure (Fig 7.3(a)). Five bond pairs and two lone pairs is found in the anion $[XeF_5]^-$, which has a pentagonal planar structure (Fig 7.3(b)), the two lone pairs residing in the axial sites. (Note that this is the other way around to how it works in a trigonal bipyramid where lone pairs go equatorial; equatorial sites are the least congested site for a trigonal bipyramid but are the most congested for a pentagonal bipyramid.) The other examples known for EX_7 and EX_5Z_2 (there are not many) adopt the same basic structure though we should note the structure (albeit the structure to be expected) of the $[IOF_6]^-$ anion which is pentagonal

Another application of Bent's rule is in rationalizing the precise structure of a compound like CF_2H_2 which is tetrahedral, but the F–C–F angle is less than the ideal tetrahedral angle because the fluorines bond through hybrids with a greater degree of p character. Deviations from idealized tetrahedral angles are calculated (and in some cases, observed) to be even more extreme in analogous lead compounds such as PbF_2Me_2.

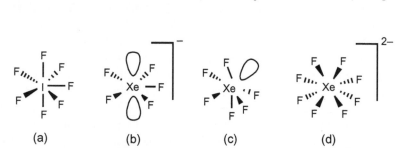

Fig. 7.3 The structures of (a) IF_7, (b) $[XeF_5]^-$, (c) XeF_6 and $[XeF_8]^{2-}$.

bipyramidal with the oxygen in one of the axial positions. For six bond pairs and one lone pair (i.e. EX_6Z), structures vary from being distorted octahedral (as is the case for XeF_6 shown in Fig. 7.3(c)) where the lone pair has some stereochemical activity to regular octahedral with a stereochemically inactive lone pair; we encountered this in the solid-state structure of BiI_3 and will return to this matter below.

Finally, the $[XeF_8]^{2-}$ anion has nine electron pairs and has a square antiprismatic structure (Fig. 7.3(d)). As a compound of Xe(VI), this is an example of a EX_8Z system and can also viewed as having a lone pair which is stereochemically inactive.

Returning to the EX_6Z examples given previously, we have noted that in this particular case the lone pair can either be stereochemically active, i.e. it occupies a space in the coordination polyhedron (which is at the heart of VSEPR rules), or it is stereochemically inactive which means that it seemingly has no influence on the coordination geometry. The regular octahedral geometry is observed, as we have seen, for bismuth in the solid-state structure of BiI_3, but is also found in the discrete anions $[SbCl_6]^{3-}$ and $[TeCl_6]^{2-}$. Yet another example is the stereochemically inactive Sn(II) lone pair in the solid-state structure of SnTe which adopts the NaCl structure. Distorted structures, such as XeF_6, are examples where the lone pair shows some stereochemical activity, and it is generally observed that this distortion is manifest as an elongated edge or expanded face of the octahedron along which or within which the lone pair is beginning to take on some stereochemical activity. Whilst VSEPR would predict that the lone pair should be stereochemically active, it offers no real guidance as to the precise structure to expect (other than perhaps a pentagonal pyramid which is not generally observed although the anion $[IOF_5]^{2-}$ with an axial lone pair and axial O is an example; recall, based on the note above for the $[IOF_6]^-$ anion, that we can consider the I–O bond in one of two ways, but either way the structure of $[IOF_5]^{2-}$ is determined by seven electron pairs). An explanation for most if not all of the observed structures can be found in terms of a second-order Jahn–Teller distortion, something we shall define, describe, and discuss for another situation in the next section, which rationalizes in a straightforward manner why and how an octahedral EX_6Z structure should distort.

Two other exceptions to VSEPR predictions are also worth noting. Thus, some bismuth and antimony penta-aryl compounds, $Sb/BiAr_5$, are observed to have square-based pyramidal geometries (we noted this earlier for the anions $[InCl_5]^{2-}$ and $[TlCl_5]^{2-}$) rather than the expected trigonal bipyramidal structure although some other examples are trigonal bipyramidal. There is probably no entirely satisfactory explanation to account for why these two different structures (and some that are intermediate) are observed except to say that for a large central atom, the energies of the two geometries are likely to be very similar and the interconversion between the two (i.e. stereochemical non-rigidity, known as Berry pseudo rotation) is known to be a low-energy process.

A second example concerns the structures of some of the Group 2 dihalides in the gas phase. Thus, whilst MgF_2 is linear as expected from VSEPR, the heavier congeners are bent and increasingly so as the group is descended [CaF_2,

Note that however we treat the I–O bonding in $[IOF_6]^-$, i.e. I=O or I^+–O^-, the structure is determined by seven pairs of electrons even though in the former case the total number of electron pairs associated with the iodine centre would be eight.

The second-order Jahn–Teller distortion for EX_6Z starts from a regular octahedron with O_h symmetry which distorts to give lower symmetry C_{2v} (elongated edge) or C_{3v} (expanded face) structures. Undistorted structures with stereochemically inactive lone pairs tend to be seen for the heavier elements, but the energy differences between undistorted and distorted structures are generally small. Thus, anions such as $[SbCl_6]^{3-}$ and $[TeCl_6]^{2-}$ can be distorted or undistorted depending on the cation present.

As an aside, $BiAr_5$ compounds are coloured (e.g. red or violet), which is unusual for organo p-block compounds. The explanation lies in the relativistic stabilization, i.e. lowering in energy, of the LUMO which has considerable 6s character such that the HOMO-LUMO gap is reduced and absorption shifts from the UV into the visible.

140°; SrF_2, 108°; BaF_2, 100°]. Two explanations have been offered. The first of these, which focuses on an ionic description of the bonding, is best exemplified by the diagrams shown in Fig. 7.4(a) and (b). Thus, if we look at the approach of an anion to a highly polarizable 2+ cation as shown in Fig. 7.4(a), a charge separation (or polarization) is set up within or across the cation as shown by the δ^+ and δ^- charges. The approach of a second anion as shown in Fig. 7.4(b) is therefore most favoured at an angle other than 180° in order to reduce unfavourable repulsions. Furthermore, the more polarizable the cation, the more bent the structure is likely to be consistent with what is observed. The second explanation centres more on a covalent bonding model in which the hybridization at the Group 2 element centre is proposed to be sd rather than sp, the former affording two equivalent hybrids which are not at 180° to each other (Fig. 7.4(c)), as opposed to two equivalent sp hybrids which are at 180°; sd hybridization is suggested to be more appropriate based on the relative energies of s *vs* p compared to s *vs* d, the latter being closer in energy in the elements for which these bent structures are observed. In fact, calculations indicate that aspects of both models, seemingly quite disparate though they appear, are required for a full description of the bonding in these species. Somewhat esoteric though gas phase dihalides may be, similar bent geometries are also found for some of the heavier Group 2 metallocenes, $[E(\eta\text{-}C_5R_5)_2]$ and alkyl compounds, ER_2.

(a)   (b)   (c)

Fig. 7.4 Ionic and covalent models for rationalizing the structures of bent Group 2 dihalides.

Before leaving VSEPR, it is worth noting that the model can accommodate simple compounds with unpaired electrons. For example, VSEPR predicts that the nitronium cation $[NO_2]^+$ is linear and that the nitrite anion $[NO_2]^-$ is bent, in the latter case because of a lone pair on the nitrogen; the observed O–N–O angles are 180° and 115° respectively. What about neutral NO_2? If we start from $[NO_2]^-$ and remove an electron from the lone pair, we are left with one electron in that orbital or domain which should repel bond pairs more than no electron but less than two. An angle greater than 115° but less than 180° is expected; the observed angle in NO_2 is 134°, consistent with this expectation. Likewise for ClO_2 *vs* $[ClO_2]^-$ for which the O–Cl–O angles are 117° and 111° respectively although this is not a dramatic difference, and calculations show that the unpaired electron in ClO_2 is most likely in a Cl–O π^*-orbital rather than in an orbital localized on chlorine so we have to be careful. We noted both of these neutral species in Chapter 4 when considering exceptions to commonly observed oxidation states.

Having provided the above overview of molecular structure according to VSEPR, it will now be useful to consider some of the periodic structural trends which are observed and to consider the various models which are used to account for them.

7.3 Trivalent compounds of Group 15

Let us start by considering the trihydrides of the Group 15 elements, EH_3. VSEPR predicts a trigonal pyramidal structure for these molecules since there is one lone pair as well as three E–H bond pairs. Moreover, as we saw previously, the interbond angles in NH_3 are less than the ideal tetrahedral angle due to lone pair repulsions being stronger than bond pair repulsions. If we look at the trihydrides of the first four members of Group 15 (BiH_3 is not stable), it is found that there is a decrease in the H–E–H angle as the group is descended as shown in Table 7.3; it is particularly marked between NH_3 and PH_3.

In fact, decreasing interbond angles on descending a group is a general phenomenon in p-block chemistry where lone pairs are present; it is also seen in the Group 16 dihydrides and in other examples discussed below. The explanation within VSEPR theory is straightforward since it is assumed that as the central atom, E, gets larger, the E–H bond pairs can be forced closer together by the lone pair before significant bond pair–bond pair repulsions arise. Moreover, the decreasing electronegativity of the element centre, as the group is descended, results in a relatively greater polarization of the bonding electron density towards the hydrogen atom. Polarization away from the central atom results in more space around that atom, which also allows the E–H bond pairs to approach more closely with a concomitant decrease in H–E–H angles. In fact, with the exception of nitrogen, the Group 15 elements are all less electronegative than hydrogen so the bonding electron density will tend to be polarized towards hydrogen, and the extent to which this is so will increase down the group.

If we now consider the trifluorides (EF_3) of Group 15, interbond angles for which are shown in Table 7.4, the first point to note is that the F–E–F angles also decrease as the group is descended just as is found for the hydrides and for much the same reasons. Note also that the interbond angle in SbF_3 is actually less than 90°, a point we will return to later, although some care must be exercised in these comparisons since SbF_3 has significant intermolecular Sb····F interactions in the solid state, unlike the lighter congeners. A second point is that the interbond angle in NF_3 is less than that in NH_3. This may, at first, seem surprising since the fluorine atom is bigger than hydrogen, but we must also recognize that the N–F bond length is greater than the N–H distance and also that fluorine is much more electronegative than hydrogen. This latter aspect is important, as the electron density will be substantially polarized towards the fluorine atoms and away from the nitrogen which will allow the N–F bond pairs to approach each other more closely at the nitrogen centre before inter-electron pair repulsions become significant.

We encounter a problem, however, when we consider PF_3, since if the above arguments for NF_3 and NH_3 are correct, we might have expected the bond angles in PF_3 to be less than those in PH_3. Clearly this is not the case here or with the arsenic analogues. A similar situation is found with H_2O vs F_2O and H_2S vs F_2S in Group 16, and there is no easy way around this difficulty based on the arguments we have employed so far. Thus, whilst VSEPR is useful for predicting the gross structural type and for rationalizing trends within a homologous series, we must

Table 7.3 H–E–H bond angles (deg) for the Group 15 trihydrides

Compound	H–E–H angle (deg)
NH_3	107.5
PH_3	93.2
AsH_3	92.1
SbH_3	91.6

Table 7.4 F–E–F bond angles (deg) for the Group 15 trifluorides

Compound	F–E–F angle (deg)
NF_3	102.3
PF_3	97.8
AsF_3	95.8
SbF_3	87.3

For the compounds NH_3, NF_3, PH_3, and PF_3, changes in the direction of bond polarity based on $\Delta\chi$ should also be considered. N–H bonds are polarized δ^- N–H δ^+, whereas the other three compounds are δ^+ N/P–H/F δ^-. This would lead to the N–H bond pairs taking up more space near the nitrogen centre and thus have the effect of widening the angles in NH_3 in relation to NF_3.

An interesting example of how this trend to smaller bond angles on descending a group affects compound stability is found with compounds of Group 14 of the general formula R_8E_8 which have the cubane structure. Thus, as the group is descended, the trend towards interbond angles of 90° results in the ring strain energy becoming progressively less.

Remember that Z in this context refers to a lone pair and not to a Z-type ligand as defined in the CBC method outlined in Section 4.1

exercise some care when comparing angles in structures with different substituents, especially where hydrogen is concerned.

Let us now further consider the trend to decreasing bond angles as Group 15 is descended. We have seen illustrations of this trend with the element hydrides and fluorides, but it is quite general for a range of substituents, and a similar effect is observed in Group 16 for compounds of the general formula EX_2 and also with respect to the inter-equatorial bond angle in EX_4Z compounds (where Z is a lone pair).

We have already seen that VSEPR provides us with an initial level of explanation, but we need also to consider the nature of the bonding (VSEPR is not a theory of electronic structure). It is sometimes stated in textbooks of inorganic chemistry that the trend towards angles of 90° reflects a tendency for the heavier elements to bond only through p orbitals (sometimes referred to as a directed valence analysis). In other words, that s-p hybridization for the heavier elements does not readily occur. This explanation, as it stands, is not easily reconciled with the fact that the energy separation between the valence s and p orbitals decreases as the groups are descended, although it is consistent with the high s-p promotion energy and particularly with inefficient s-p hybridization resulting from the different sizes of the ns and np orbitals for the heavier elements (see Chapter 4).

It would therefore seem that bonding through pure p orbitals might be an appropriate description for the heavier elements, and that in, for example, triphenyl bismuth, $BiPh_3$, with C–Bi–C angles of 94°, the bismuth–carbon bonds are formed by overlap involving bismuth p orbitals, and that the bismuth lone pair resides in a pure s orbital. This latter point is also consistent with the observed decrease in basicity of Group 15 EX_3 compounds as the group is descended. Thus, whilst PPh_3 readily forms complexes by acting as a Lewis base, the analogous chemistry for $BiPh_3$ is much less extensive due to s orbitals with poor directional properties not forming strong bonds which, coupled with the radial contraction of the heavier element s orbitals, leads generally to weak bonds and reduced basicity.

The argument above is certainly one way of looking at things and is fairly satisfactory at explaining many features, although it does not readily account for why interbond angles of less than 90° should be observed as found, for example, in SbF_3 (and TeH_2). An alternative approach to this problem, and ultimately a more satisfactory one, is to consider the nature of the molecular orbitals and how these can be used to rationalize the shapes of molecules.

Let us start by asking the question of why EX_3Z molecules are pyramidal from a molecular orbital point of view. In so doing we can start by considering a more symmetrical trigonal planar reference structure and constructing the molecular orbitals for this geometry. Fig. 7.5 shows a qualitative molecular orbital energy level diagram for planar EH_3 and how the orbitals change in energy as the geometry is changed to trigonal pyramidal. This is known as a **Walsh diagram**; details of how it is constructed, and the symmetry arguments involved, are beyond the scope of this book but can be found in the books on molecular orbital theory listed in the bibliography.

The important points to note are as follows. For a planar, 8-electron species, i.e. EX_3Z, the HOMO is the non-bonding a_2'' orbital in D_{3h} symmetry and the

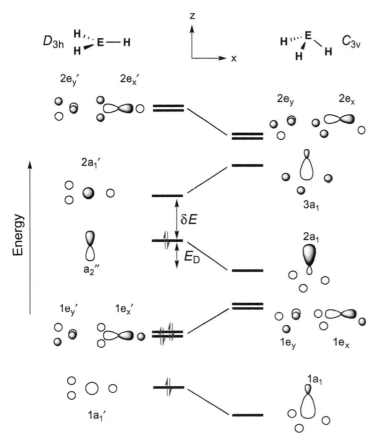

Fig. 7.5 An orbital interaction diagram or Walsh diagram for EH$_3$. For a 6-electron species, the electrons are shown in blue for the planar reference structure; the additional pair for an 8-electron structure are shown in red.

LUMO is the 2a$_1$' σ-antibonding orbital. As the symmetry is lowered from planar to trigonal pyramidal, i.e. D_{3h} to C_{3v}, these two orbitals take on the same symmetry, a$_1$ in C_{3v}, and are therefore allowed to mix which results in a lowering of the energy of the HOMO (now 2a$_1$) since it becomes slightly bonding, and a raising in energy of the LUMO (now 3a$_1$) as it becomes more antibonding. This type of distortion is known as a **second-order Jahn–Teller distortion**, and it is the lowering in energy of the HOMO which drives the distortion and results in 8-electron EX$_3$Z species having a trigonal pyramidal geometry rather than being trigonal planar. Note that in a 6-electron species such as BH$_3$, the doubly degenerate HOMO is lower in energy for the trigonal planar structure compared to the trigonal pyramidal structure consistent with the observed geometry for this molecule.

We can take this argument a little further by considering the extent to which a distortion from planar to pyramidal will occur. In other words, what factors determine how pyramidal the molecule will become? In fact, this is governed largely by the energy separation between the HOMO and LUMO in the planar reference structure (the HOMO–LUMO gap labelled δE in Fig. 7.5). In situations where the energy separation is small, the orbitals can mix more effectively and the corresponding distortion to pyramidal (indicated as E_D in Fig. 7.5) will be large, whereas for a large energy gap only small distortions will be observed.

If we now consider the energy levels for the Group 15 EX_3Z species, the energy gap (δE) will decrease as the group is descended for two reasons. Firstly, the a_2'' HOMO will rise in energy as the element centre becomes less electronegative, and secondly, the $2a_1'$ LUMO will fall in energy as the central atom becomes larger since the orbitals are getting bigger and more diffuse such that overlap is less efficient and the orbital is not so antibonding. We can also account for why it is that for a given element, E, the molecule will become more pyramidal as the electronegativity of X increases. This is due to a greater separation in energy between the orbitals of E and X which will lower the energy of $2a_1'$ and decrease δE.

Not only does this model rationalize why pyramidality increases down the group, it also accounts for a related observation that pyramidal inversion energy barriers increase as the group is descended. In NH_3, the barrier to inversion at the nitrogen centre is about 25 kJ mol^{-1} whereas for PH_3 it is much larger at around 140 kJ mol^{-1}. This is directly related to the increasing pyramidality, since, as a molecule becomes more pyramidal, the energy difference between $2a_1$ in C_{3v} and a_2'' in D_{3h} becomes larger. In fact, this energy difference can be taken as a measure of the activation energy for inversion through a planar transition state.

Note that in all of these arguments, no reference is made to a more traditional hybridization model, and also that there is nothing particularly special about an interbond angle of 90°. There is no reason why the distortion should not continue and lead to angles less than 90° if this results in a further lowering of the energy of the HOMO.

Another trend which can be rationalized along similar lines concerns the increasing pyramidality found for the higher homologues of alkenes, $R_2E=ER_2$ (R = C to Sn). Thus, whilst alkenes are generally planar, Fig. 7.6(a), the so-called disilenes, digermenes, and distannenes increasingly adopt a *trans* bent geometry where each E centre is trigonal pyramidal rather than trigonal planar, Fig. 7.6(b). A molecular orbital scheme and an associated second-order Jahn–Teller explanation can be employed to account for this trend, but also relevant are the energetics of the hypothetical fragments R_2E. Thus, if the so-called singlet–triplet energy difference is small, as it is for carbon, two triplet fragments can combine to give a planar structure, as shown in Fig. 7.6(c), whereas for the larger singlet–triplet

These ideas based on Fig. 7.5 can also be used to rationalize why the CH_3 radical is planar whereas CF_3 and related silicon radicals are pyramidal.

We should note that whilst pyramidal inversion energy barriers through a trigonal planar geometry do increase down the group in Group 15, for the heavier congeners an alternative process involving a T-shaped transition state has been found to be lower in energy.

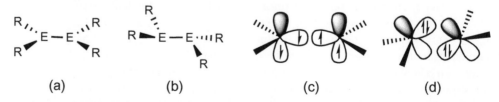

(a)   (b)   (c)   (d)

Fig. 7.6 Planar (a) and *trans* bent (b) forms of $R_2E=ER_2$ and representations of the overlap of two triplet (c) and two singlet (d) R_2E fragments. A triplet state is one where there are two unpaired electrons, whereas these electrons are paired in a singlet state.

energy differences which occur as the group is descended (due to increasing s-p promotion energies, see Fig. 2.7), two singlet fragments combine to give the *trans* bent structure, Fig. 7.6(d).

More recently, several examples of disilynes (RSi≡SiR) and heavier Group 14 congeners have been prepared, and these have been found to have *trans* bent structures as opposed to the linear structure found for alkynes (RC≡CR) as illustrated in Fig. 7.7. An explanation for the non-linearity of these heavier Group 14 species parallels that seen for pyramidalization in disilenes etc.

The molecular orbital arguments presented above are quite general and can be used to rationalize most, if not all, of the structural trends observed in p-block chemistry. As is so often the case in chemistry, two apparently different theories or approaches are complementary rather than contradictory. The molecular orbital method is useful for understanding electronic structure, but VSEPR is preferable in view of its simplicity, for quickly predicting structure.

7.4 The Zintl principle

As a final section on structure, we will consider the **Zintl principle** which is helpful in identifying structural similarities between seemingly disparate compounds. Another name for this idea is the **isoelectronic principle** which is much more descriptive since it indicates that the key to structural similarity is to be found in the number of electrons (specifically valence electrons). Thus, if a set of compounds has the same overall number of valence electrons they are likely to have similar structures provided that the elements do not differ widely in electronegativity or polarizability.

As a first example, we can consider polymeric compounds which have the cubic diamond structure. These include the so-called III–V (or 13–15) compounds boron nitride (BN), aluminium phosphide (AlP), and gallium arsenide (GaAs), the II–VI (or 12–16) compounds zinc selenide (ZnSe) and cadmiun telluride (CdTe), and the I–VII (or 11–17) compounds copper(I) bromide (CuBr) and silver iodide (AgI), all of which have four bond pairs per atom and no lone pairs.

For more ionic systems, an example is the structure of magnesium diboride, MgB_2. In this compound the magnesium is present as Mg^{2+}, the two electrons lost having been transferred to the boron atoms. Since there are two borons per magnesium, each boron is present as B^- which is isoelectronic with carbon, and the structure comprises planar, hexagonal sheets of boron atoms isostructural with the sheets in graphite (Fig. 3.4(b)); the Mg^{2+} cations are arranged between the boron layers.

Similarly, in lithium arsenide, LiAs, the lithium is present as Li^+ and the arsenic is present as As^-, which is isoelectronic with neutral Se. This is reflected in the arrangement of arsenic atoms which is isostructural with elemental grey selenium (Fig. 3.6(c)), i.e. helical. In calcium disilicide, $CaSi_2$, the silicons are present as Si^-, which is isoelectronic with P, and the silicon atoms are arranged in a manner isostructural with orthorhombic black phosphorus (Fig. 3.5(c)), i.e. puckered hexagonal sheets. In K_4Si_4, however, the silicons are present as Si_4^{4-} tetrahedra analogous to tetrahedral P_4 or white phosphorus.

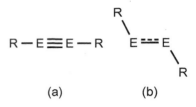

Fig. 7.7 Linear (a) and *trans* bent (b) forms of RE≡ER.

The general rule is that if $E_{\sigma+\pi} > 2\Delta E_{S-T}$ (where ΔE_{S-T} is the singlet–triplet energy difference) a classical planar structure is obtained, whereas if $E_{\sigma+\pi} < 2\Delta E_{S-T}$ a *trans* bent structure is observed. If $E_{\sigma+\pi} < E_{S-T}$, singlet monomers with no E–E bonding will result.

A question which often arises is why a species such as $SnCl_2$ forms a polymeric, chlorine-bridged structure (Chapter 5) rather than a distannene with a Sn–Sn bond (i.e. $Cl_2Sn-SnCl_2$). This reflects the greater availability of the chlorine lone pair for bonding as opposed to the tin lone pair, i.e. Sn–Cl bonds are stronger than Sn–Sn bonds.

Space does not permit a fuller discussion of the molecular orbital theory employed in the section. For this and a more in-depth treatment of second-order Jahn–Teller effects, the reader is referred to Albright, Burdett, and Whangbo (1985) and Gimarc (1979).

13–15, for example, clearly refers to the respective groups in the periodic table. Many of these compounds are, however, still referred to using the Roman numeral designation III–V derived from the old group numbering scheme (i.e. main Groups III and V).

As a further example, we may note that the thallium atoms in the structure of NaTl are present as Tl⁻ and are arranged in a diamond-like, three-dimensional network. Many ternary compounds can also be rationalized according to these ideas. In terms of the van Arkel–Ketelaar triangle discussed in Chapter 4, the ionic materials described above would lie in a band that is close to and parallel with the metallic-ionic edge.

Finally, we can note that in calcium carbide, CaC_2, the C_2 unit is present as a discrete C_2^{2-} ion isostructural with N_2 which illustrates the prevalence of π-bonding amongst the 2p elements.

Exercises

1. Use VSEPR to predict the structures of the following molecules: $[PF_4]^-$, $PhI(OH)_2$, SO_2Cl_2, $[PO_4]^{3-}$, SNF_3, IF_5, XeO_3, XeO_2F_2, I_2Cl_6.

2. Predict possible structures for the p-block component of the following compounds: Li_2S_2, $BaSn$, Cs_2NaAs_7.

8 Theories and models: scope and limitations

We will conclude this book with a new chapter which was not part of the first edition. Here, we shall focus on some more recent developments in terms of bonding, say a little more about the involvement of d orbitals, and finish with a section on the application of some of the various theories and models that we have encountered in earlier chapters.

8.1 Bonding

Let us first consider the nature of bonding itself. This is a very big topic, of course, and we can hardly do it justice here, but in Chapter 4 in particular we looked at the van Arkel–Ketelaar triangle and highlighted how it can be used to understand elements and compounds in terms of three types of bonding: metallic, ionic, and covalent. We also saw how this approach could be taken a step further with the element tetrahedron to encompass van der Waals interactions. Straightforward though this might all seem, reference was made to the argument of Allen and Burdett who had questioned the value of the term 'metallic bond' since, in their view, this is just another form of covalent bonding.

The basis of the argument that Allen and Burdett make is as follows. Ionic bonds or interactions can be characterized according to a basic expression such as $-e^2/r + A/r^{12}$ (e is the unit of charge, r is the distance between the ions and A is a constant). This is the essence of equations such as the Born–Landé equation (Eqn. 5.1) for calculating lattice energies where an electrostatic Coulombic attraction ($-e^2/r$ in the case of singly charged ions) is balanced by a repulsive term (A/r^{12}). The strength of van der Waals interactions can be characterized using an expression of the form $-B/r^6 + A/r^{12}$ which also balances attractive (the former) and repulsive (the latter) terms (r is again distance and B and A are constants; the A/r^{12} term is the same for both ionic and van der Waals bonding, and is approximate, and we should recognize that the attractive term $-B/r^6$ is also present in ionically bonded systems albeit very much less important than the $-e^2/r$ term). For covalent bonding, the strength of the interaction is determined by evaluating the overlap between the relevant orbitals, the mathematical expression for which we will not be concerned with here.

The point these authors then make is that any evaluation of the strength of metallic bonding is determined in the same way as for covalent bonding, and therefore that the term 'metallic bond' is unnecessary and redundant. Does all this really matter? It depends a little on what you are trying to explain. Inasmuch as metallic bonding involves the sharing of electrons, it is covalent bonding by definition, albeit bonding which is delocalized in three dimensions rather than the more localized bonding arrangements appropriate for molecular or polymeric compounds. As we said before, however, the metallic state is associated with some important properties (whether in metals themselves, intermetallics and alloys, or even those oxides, sulfides and other compound materials that exhibit metallic conductivity), and it is certainly useful to recognize this in terms, for example, of the van Arkel–Ketelaar triangle or element tetrahedron.

Herein lies a distinction that is worth making. On the one hand we can consider the fundamentals of bonding according to Allen and Burdett. On the other hand, it is often very useful to focus on properties, and reference to metallic bonding can be seen as useful in that regard. This should remind us a little of the arguments about the structure of the periodic table which we touched on in Chapter 1. Should it be based on atomic ground state electronic configurations or more on the chemical/physical properties of the elements; what is the right place for He? There is no 'correct' answer.

Part of the reason for the detailed discussion presented above is to provide a preface for some interesting new discussions around bonding, particularly in terms of the properties of certain materials. Thus, recent (2019) work by Wuttig has introduced the term and idea of ***metavalent bonding*** and makes the claim that this should be considered as a new type of bonding distinct from covalent, metallic, or ionic. Wuttig's proposal focuses principally on the properties of compounds such as GeSe, GeTe, SnSe, SnTe, PbSe, and PbTe (i.e. 1:1 compounds of the heavier Group 14 elements in their +2 oxidation state with selenium or tellurium) although it is by no means restricted to this group. Wuttig refers to these materials as 'metavalent solids' or 'incipient metals', and focuses on their properties which are distinct not just from metals and from covalently bonded solids (which, in the case of covalently bonded solids, obey the 8 − N rule in terms of the number of bonds they form) but also from metalloid materials such as those mentioned in Chapter 4 including GaAs, InSb, SiC and AlP, GaP, InP, all of which adopt the diamond structure with tetrahedrally coordinated atoms (and do obey the 8 − N rule). Properties such as high electrical conductivity, particular thermoelectric properties, and a number of other features which we will not list here, are considered sufficiently distinct from those of metals, metalloids, and covalent solids to warrant the introduction of this new category.

Whether or not these unusual properties justify the introduction of the new term 'metavalent' bonding remains open to debate. No doubt Allen and Burdett would describe the bonding in these compounds as just another manifestation of covalent bonding, but if we give primacy to chemical or physical properties, perhaps this definition offers a new insight. Either way, the work of Wuttig has been summarized in the plot shown in Fig. 8.1. The relationship between this plot and the van Arkel–Ketelaar triangle is clear to see, and indeed the triangle

Another argument we might employ to retain the term metallic bonding is the following. In molecular covalent species, most compounds are in their electronic ground state whereas compounds which exhibit metallic bonding have a partially filled band, and at temperatures above absolute zero, electrons populate some of the higher energy levels. Metals are therefore in an excited electronic state, and many of their particular properties are a direct result of this feature.

More details on metavalent bonding can be found in Wuttig et al., *Adv. Mater.*, 2019, **31**, 1806280.

The 14–16 materials listed here have structures based on the NaCl structure, or a distorted variant thereof, wherein the Group 14 element lone pair is either stereochemically inactive or somewhat active respectively. The Pb structures are undistorted whilst the Ge structures show the most degree of distortion, the explanation for which can be cast in terms of a Peierls distortion from a symmetric structure to a less symmetric one.

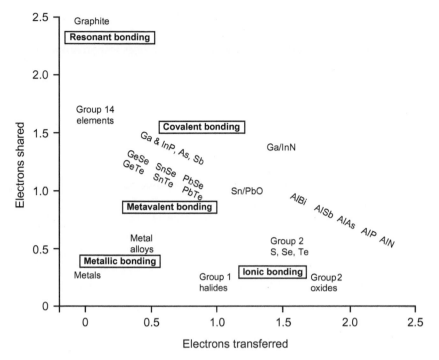

Fig. 8.1 A compound plot showing regions associated with the bonding regimes metallic, covalent, ionic, metavalent, and resonant.

Details of how the 'electrons shared' and 'electrons transferred' are quantified are to be found in the reference quoted to the original work.

in Fig. 8.1 defined by metallic bonding, covalent bonding, and ionic bonding is basically the van Arkel–Ketelaar triangle albeit in a different orientation to that shown in Fig. 4.6. We see that the region identified as comprising compounds that exhibit metavalent bonding largely coincides with the metalloid region highlighted in Fig. 4.8, but Wuttig and others distinguish between metalloids and metavalent compounds in terms of their structures, properties, and the precise details of their bonding, which we will not dwell on here.

Another type of bonding highlighted by Wuttig and labelled **_resonant bonding_** is also shown in Fig. 8.1. This applies to graphite and related graphitic materials with extensive π-delocalization found in association with a more localized sp^2 framework.

8.2 d orbitals again

In Chapter 6 we considered the matter of d orbitals in p-block chemistry in terms of their involvement in hybridization in species such as PF_5 (dsp^3) and also as possible acceptor orbitals to account for the formation of complexes such as $[SiF_4(py)_2]$. As we stated at the time, a large part of the reason for invoking the use of d orbitals was to retain bonding descriptions that featured two-centre, two-electron bonds and to account for why many so-called hypervalent species were in apparent violation of the octet rule. We showed, however, that for species such as I_3^- and XeF_2, a bonding model involving three-centre, four-electron bonds is an alternative to using hybrids containing d orbitals, and also that the σ*-acceptor model provides an alternative to employing vacant d orbitals when it comes to explaining Lewis acidity.

As a further example which illustrates that d orbitals are unnecessary when seeking to account for the bonding in hypervalent species, we can consider SF_6. SF_6 has a regular octahedral structure (point group O_h), and a molecular orbital energy level diagram for it is shown in Fig. 8.2. The key point is that the a_{1g} and t_{1u} symmetry adapted orbitals for the six fluorine atoms find a match with the sulfur s and p orbitals (a_{1g} and t_{1u} respectively in O_h symmetry) whereas, in the absence of d orbitals, the e_g set do not. The molecular orbitals shown in black in Fig. 8.2 therefore comprise four bonding ($1a_{1g}$ and $1t_{1u}$), four antibonding ($2a_{1g}$ and $2t_{1u}$), and two non-bonding ($1e_g$) orbitals. The available electrons fill the four bonding and two non-bonding giving twelve in total, but only eight are in bonding orbitals. The remaining four are localized on the fluorines with no contribution from the sulfur. This is entirely analogous to the argument advanced in Chapter 6 for XeF_2 wherein the non-bonding pair is also localized only on the fluorine atoms with no contribution from Xe.

Two of the d orbitals on sulfur do have e_g symmetry in the O_h point group, however. These are the $d_{x^2-y^2}$ and d_{z^2} orbitals. If these orbitals are used, a further two bonding orbitals are formed (shown in red in Fig. 8.2) as well as two further antibonding orbitals (not shown), i.e. what was non-bonding becomes bonding. Overall, this would now result in six bonding orbitals and a description of SF_6 as a

The terms a_{1g}, t_{1u} and e_g are derived from group theory, but the details of how this diagram is constructed need not detain us here (more detail can be found in any of the texts on bonding listed in the bibliography).

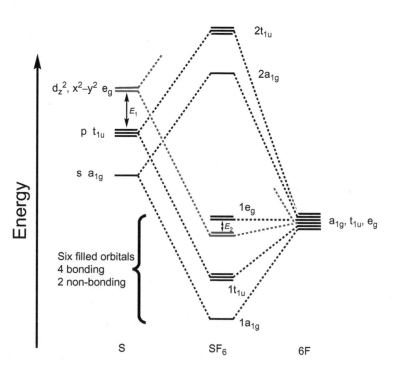

Fig. 8.2 A molecular orbital energy level diagram for SF_6.

12-electron species would be valid (and we could sensibly talk of d^2sp^3 hybridiza-tion and six two-centre, two-electron bonds). The question, of course, is: are the d orbitals involved? From a symmetry consideration there is no problem, so it is all about energy. If the d orbitals lie just above the p orbitals (i.e. if the energy gap E_1 in Fig. 8.2 is small) then their inclusion is reasonable, but if they lie far above (if E_1 is large), it is not. Calculations indicate the latter and that there is no significant d orbital involvement, i.e. E_1 in Fig. 8.2 is large, and as a result, E_2, which is a meas-ure of the extent to which the non-bonding e_g orbitals become slightly bonding due to d orbital involvement, is insignificant. The point we have therefore made is that from a molecular orbital perspective, d orbital involvement is neither sig-nificant nor required in this or in any similar species, and in terms of the number of electrons associated with the sulfur centre, the octet rule is not violated.

8.3 Models and theories

It will be clear from the discussion in many if not all of the previous chapters that a variety of theories and models play a vital rôle in helping us to understand all the various aspects of s- and p-block chemistry with which this book is concerned. This is apparent from as early as the beginnings of Chapter 1 where we looked at wave equations and orbitals and made the point that the very commonly used cartoon-like representations of orbitals shown in Fig. 1.4 are extremely useful in understanding many aspects of bonding and structure despite the known inac-curacies of these types of pictures. Does this matter? No, it does not, as long as we appreciate the limits of our models and recognize the domains in which they can be applied and those in which they cannot. We shall explore this in a little more detail in this section.

Let us consider a commonly used model, this time from organic chemistry. Curly arrows are used to represent the movement of pairs of electrons when considering organic reaction mechanisms. No case is made that electron pairs actually behave in this manner during the course of a reaction, but curly arrows are an extremely useful way of understanding mechanistic pathways. Indeed, this is why benzene (and arenes generally) are usually represented as shown in Fig. 8.3(a) since this makes the movement of electron pairs using curly arrows much easier than if we were to use the image shown in Fig. 8.3(b). This brings us directly to an important point. We can and do use the picture of benzene shown

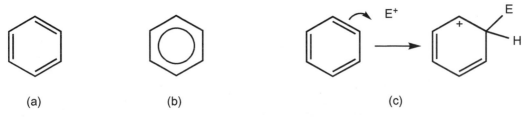

(a)   (b)   (c)

Fig. 8.3 Two different representations of benzene (a) and (b), and the first step of electrophilic aromatic substitution (c).

in Fig. 8.3(a) because it helps us understand mechanistic pathways even though we know that this cyclohexatriene rendition of the structure with alternate single and double bonds is not a satisfactory description if, for example, we want to better represent the delocalized electronic structure in which case Fig. 8.3(b) is more suitable. It is therefore all about what we are trying to explain; both pictures or representations are approximate, but both nevertheless have value. What we must not do is mix them up. Fig. 8.3(a) is not a good representation of the de-localized electronic structure of benzene, but Fig. 8.3(b) is of little use when it comes to explaining mechanisms such as electrophilic aromatic substitution, the first step of which is shown in Fig. 8.3(c).

To choose another example, let us return to VSEPR which we considered in some detail in Chapter 7. The rules of VSEPR are predicated on the mutual repul-sion of electron pair domains in the valence shell of the central atom of a mol-ecule. It works extremely well for a large range of examples and is a very useful heuristic. However, VSEPR is based on electron pairs being localized between pairs of atoms or located solely on the central atom in the case of a lone or non-bonding pair. As we saw in Chapter 6, this is not generally a very good descrip-tion of bonding and the electronic structure of molecules. We saw, for example, that in the I_3^- anion or in XeF_2, the linear tri-atomic unit is better described using a three-centre, four-electron bonding model rather than considering each of the two bonds to a be a localized pair. This in turn led us to question whether the octet rule was violated or not in such compounds; we have just examined a similar argument with SF_6.

The point here is that VSEPR works very well within the domain for which it was developed, which is to rationalize the structure of (predominantly) p-block element compounds. What it is not is a theory or model of electronic structure, and to therefore seek to apply it in that context is misguided. Consider methane, CH_4. VSEPR predicts a regular tetrahedral structure based on there being four equivalent C–H bonding pairs around the carbon. This is shown in Fig. 8.4(a). Both structural and spectroscopic data confirm that all four C–H bonds are equivalent, and we can account for this by assuming that the valence 2s and three 2p orbitals of carbon are hybridized to form four equivalent sp^3 hybrids which are shown schematically in Fig. 8.4(b). This hybridization model provides a convenient description of the bonding in terms of localized two-centre, two-electron bonds, which is implicit, although not explicitly required, in VSEPR. So far, so good, but if we look at the photoelectron spectrum of methane which

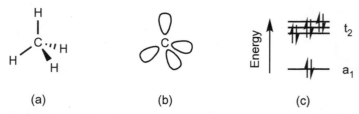

(a)   (b)   (c)

Fig. 8.4 (a) The structure of methane, CH_4, (b) four sp^3 hybrid orbitals, and (c) part of the molecular orbital energy level diagram for CH_4.

provides us with a direct measure of the energies of the bonding electrons, we observe that there are two sets of bonding orbitals, one lower in energy than the other. This can be rationalized in a straightforward manner by molecular orbital theory (which we will not consider in detail here), and a part of the molecular orbital energy level diagram for methane is shown in Fig. 8.4(c) which reveals a single bonding orbital at lower energy and a triply degenerate set of bonding orbitals at higher energy. All four molecular orbitals are delocalized across all five atoms, but the average is one bonding pair per C–H bond, so the various descriptions are not contradictory in that sense.

What this example helps us to appreciate is that asking which model is right or correct is the wrong sort of question and one that can all too easily lead to difficulties. Both are approximations, and the point we should recognize is that it is perfectly acceptable to make use of approximate models. Chemists do it all the time, and with good reason, for these simple models allow us to make sense of a vast body of data or facts. Where we have to be careful is to make sure that we do not use any particular model outside of the domain or confines for which it is appropriate. In the example of methane, VSEPR and hybrid orbitals are fine in terms of rationalizing molecular structure, but they are not fine if we wish to understand details of the electronic structure and some spectroscopic data; in this case we need molecular orbital theory.

Let us briefly consider another example. Recall how we represented the bonding between phosphorus and oxygen in $POCl_3$ in Fig. 4.2 in Chapter 4. The most commonly encountered form is as a double bond, i.e. P=O (Fig. 8.5). From detailed electronic structure calculations, we know that this is not quite right and is at best a poor characterization of the phosphorus–oxygen bond, but it is useful nevertheless in understanding everything from the reactivity of phosphine oxides to the structural chemistry of phosphates. If it does not quite work when it comes to the detailed electronic structure of such species, that is fine as long as we recognize it.

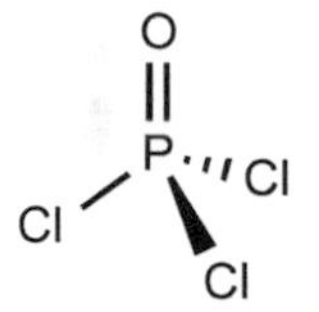

A question often asked is that surely there must be one model or theory or explanation that works best in all cases. Well, up to a point. For any molecule, we can use quantum-chemical calculations to determine the structure of lowest energy, and modern computational methods with access to high-performance computing hardware allow us to do just that. At a certain level that is exactly what we wish to do, but that does not mean that simple models, approximate and limited though they are, are without value especially in a teaching context. To seek to explain all the varied chemistry illustrated in this text with reference to statements such as detailed calculations account for this or that structure or this or that reaction pathway would have essentially no pedagogical value. Teachers and practicing chemists alike need approximate models, and that is why they are important.

Does this mean that pretty much anything goes as a model? No, not really. Without getting too philosophical, what we should expect of our models is that they meet the requirements of any scientific model or theory. They must account for what we observe, but they should also have predictive power, i.e. can we use the model to make a prediction (testable by experiment or observation) about

Some will argue that molecular orbital theory is inconsistent with the presence of localized electron pairs and that the basis for VSEPR is therefore flawed. Others, notably Gillespie, point to the fact that whilst the total electron density, ρ, may not reveal evidence of bond pair localization, the so-called negative Laplacian or second derivative of the total electron density, $-\nabla^2\rho$, does.

In a photoelectron spectrum, UV radiation of known frequency and hence energy is used to eject (ionize) valence electrons from the molecule being studied. The kinetic energy of the ejected electrons is then measured, and the difference between the energy of the UV radiation and the kinetic energy of the electron is a direct measure of the binding or ionization energy of the electron.

Fig. 8.5 One representation of the structure of $POCl_3$.

the structure or some other feature of a new molecule or material not yet determined? This is a fundamental axiom of the scientific method. If the answer is that they do, these models have value.

To conclude, chemistry perhaps more than any other discipline makes use of a multitude of approximate models; rules of thumb, if you like, or heuristics. Sometimes it may even seem to be a bit hand-wavy, but whilst this may come across to some as a deplorable lack of rigour, most chemists would argue that it is essential, particularly in a teaching context. The key is to recognize the limitations of the models and to not seek to apply them beyond these limits. In a 2019 review entitled 'Chemical Bonding and Bonding Models in Main-Group Chemistry' (*Chem. Rev.*, 2019, **119**, 8781), Frenking and Schwerdtfeger and co-workers offer the following statement: *Chemical bonding models are not right or wrong, they are more or less useful.*

Glossary

Allotrope This term refers to and distinguishes different structural forms of the same element; for example, diamond and graphite.

Alternation effect The observed trend in the stability of the highest oxidation states encountered on descending a Group in the p-block; for example, in Group 15: P(V), As(III), Sb(V), Bi(III).

Amorphous A solid in which atoms or molecules have no long-range order.

Angular wavefunction The angular part of the wavefunction varies as a function of the angles in a spherical coordinate system and thus determines the shape of the orbital.

Angular node A node in the angular part of the wavefunction.

Atomic radius A somewhat arbitrarily defined radius of an atom. In metals, this is half the distance between nearest neighbours.

Aufbau Principle The principle which states that electrons are added to the lowest-energy orbital available.

Azimuthal or angular momentum quantum number This quantum number describes the sub-shell of the electron and is associated with the labels s, p, d, f.

Band theory Molecular orbital theory applied to extended solids.

Bohr radius The most probable distance from the nucleus for an electron in a 1s orbital in a hydrogen atom, equal to 0.529 Å.

Bond valence A concept employed in solid-state chemistry to assign a valence (actually an oxidation state) to an element centre in an extended macromolecular solid.

Born-Haber cycle A method of determining lattice energies from other data such as atomization energies, ionization energies, electron affinities, and heats of formation.

Boundary conditions In the context of electrons in atoms, the fact that electrons are constrained by their attraction to the positively charged nucleus.

Brønsted-Lowry A definition of acids and bases where an acid is a proton (H^+) donor and a base is a proton acceptor.

Catenation Compounds which contain element-element bonds where the atoms are the same; for example, the P-P bonds in black phosphorus.

Conjugate acid The acid derived from protonating a base.

Conjugate base The base derived from deprotonating an acid.

Coordinate bond Another term for a dative covalent bond (see below).

Coordination number The number of atoms directly bonded to the element in question.

Core electrons Electrons in lower-energy orbitals than the valence electrons which are not directly involved in bonding but which exert an indirect effect through shielding.

Covalence For a given element, the number of covalent bonds formed by that element in the molecule under consideration.

Covalent radius The covalent radius of an atom is defined as half the distance of a typical element-element single bond. Double and triple bond covalent radii can be similarly defined.

Crystalline A solid in which atoms or molecules have long-range translational order.

Dative covalent bond A covalent bond between two elements in which the two electrons formally come from only one of the atoms.

Degenerate Degenerate states have the same energy.

Diagonal relationship The relationship between pairs of elements in the s- and p-block related diagonally in terms of their position in the periodic table; for example, B and Si.

Effective nuclear charge The actual nuclear charge felt by an electron after electron shielding has been considered.

Electron affinity The energy associated with adding an electron to an atom (or ion).

Electron deficient Compounds for which the number of valence electrons is insufficient to fully fill the valence orbitals; for example, BH_3.

Electron precise Compounds for which the number of valence electrons is sufficient to fully fill the valence orbitals; for example, CH_4.

Electronegative A description of elements which have a large electronegativity.

Electronegativity Defined as 'the ability of an atom to attract electron density towards itself in a molecule'.

Electropositive A description of elements which have a small electronegativity.

Element tetrahedron A graphical means of visualizing the relationship between covalent, metallic, ionic, and van der Waals bonding in compounds.

Exchange energy An energy derived from quantum mechanics which relates to the energies associated with particular arrangements of electrons in degenerate orbitals which is the basis of Hund's First Rule.

Excited state An energy state higher in energy than the ground state.

Ground state The lowest energy state.

Hess's Law The law which states that an energy change in, for example, a Born-Haber cycle is independent of the route taken to determine it.

HOMO Highest Occupied Molecular Orbital.

Hund's First Rule The rule which states that for a set of degenerate orbitals, electrons are added such that they occupy separate orbitals and with their spins aligned (i.e. not spin paired).

Hypervalence A term used to describe compounds in which the total number of valence electrons around the central atoms exceeds or appears to exceed eight.

Inert pair effect The observation that many of the heavier p-block elements, particularly the 6p elements, often exhibit an oxidation state in their compounds two less than the Group maximum.

Ionic radius The radius of an ion in an ionic solid derived in a self-consistent manner from a range of interionic distances determined for a range of ionic compounds.

Ionization energy The energy required to remove an electron from an atom (or ion).

Isomorphous A term meaning 'has the same crystal structure as'.

Isostructural A term meaning 'has the same molecular structure as'.

Lattice energy Defined as the energy required to convert one mole of an ionic solid into one mole of gaseous ions.

Lewis acid An electron pair acceptor.

Lewis base An electron pair donor.

Lobe Displacement between nodes.

LUMO Lowest Unoccupied Molecular Orbital.

Lux-Flood A definition of acidity in which an acid is an oxide acceptor and a base is an oxide donor.

Madelung Rule. The rule which states that orbitals with a lower value of the quantum numbers $n + l$ are filled first.

Magnetic quantum number This quantum number describes the particular orbital in a given sub-shell.

Metavalent bonding A term which defines a type of bonding which characterizes compounds in which electrons are shared but which have properties distinct from normal covalent solids and metalloids.

Modification See polymorphism.

Node A point of zero amplitude.

Orbital An allowed energy state of an electron in an atom.

Oxidation state Defined for a particular element as a formal charge which results from the bonded atoms being partitioned according to their respective electronegativities.

Pauli Exclusion Principle The principle which states that no two electrons in an atom can have the same set of four quantum numbers.

Penetration The extent to which electrons in one orbital can penetrate another, lower–energy orbital and thereby feel a greater effective nuclear charge.

Polymorphism A term used to distinguish between different crystalline modifications of the same molecular form such as, in the case of elements, the orthorhombic and monoclinic forms of S_8.

Principal quantum number This quantum number describes the electron shell or energy level.

Principle of maximum hardness A principle derived from the theory of hard and soft acids and bases in which chemical reactions proceed to give products of maximum hardness which equates to species in which electrons are in the lowest-energy orbitals.

Probability amplitude The amplitude of the electron wave in a particular orbital.

Probability density The square of the amplitude of the electron wave in a particular orbital.

Promotion Energy The energy required to promote an electron from an orbital of lower energy to one of higher energy.

Prototropic tautomerism Two molecules which differ only in the location of a hydrogen atom.

Quantization In the context of electrons in atoms, the fact that only discrete energies are allowed.

Quantum number A number or numbers which give allowed solutions to the wave equation.

Radial node A node in the radial part of the wavefunction.

Radial probability (distribution) function The function which determines the most probable distance at which the electron will be found from the nucleus.

Radial wavefunction This part of the wavefunction reveals how the wavefunction varies with distance, r, from the nucleus, i.e. the effective size of the orbital.

Radius ratio rule A rule which allows an approximate determination to be made as to whether a cation (usually) will occupy a tetrahedral or an octahedral hole in a close-packed array of anions.

Relativistic effects A term used to describe the consequences of the theory of special relativity on the mass of electrons moving at close to the speed of light and the corresponding effect on orbital radii.

Resonant bonding A term used to describe the extended π-delocalization of electrons in materials such as graphite.

Secondary Bonding A term used to describe the close, non-covalent interactions between atoms often seen in compounds of the heavier p-block elements.

Second-order Jahn-Teller distortion A distortion from a more symmetrical structure to a less symmetrical structure driven by changes in the energies of molecular orbitals.

Shielding For atoms or ions with more than one electron, the effect that electrons have on each other in terms of shielding and hence reducing the nuclear charge felt by the outermost or highest-energy electrons.

Slater's rules An approximate method of calculating a shielding constant in order to calculate an effective nuclear charge.

Spin In the case of an electron, the angular momentum associated with the electron itself.

Spin paired Two electrons which have opposite spins.

Valence For a given element, the number of valence electrons used in bonding.

Valence electrons The electrons in the highest-energy orbitals and those involved in bonding and therefore associated with the chemistry of the atom.

van Arkel-Ketelaar triangle A graphical means of visualizing the relationship between covalent, metallic, and ionic bonding in compounds.

Wave equation In this context, an equation which describes the properties of electrons in atoms.

Wavefunction One of the solutions to the wave equation.

Zintl principle The principle that isoelectronic species have related structures in solid-state chemistry; for example, B_n^- sheets being isoelectronic and hence isostructural with C_n sheets in graphite.

Bibliography

General inorganic chemistry texts

1. Weller, M. T., Overton, T. L., Rourke, J. P., Armstrong, F. A. (2018). *Inorganic Chemistry* (7th edn), Oxford University Press.

2. Housecroft, C. E., Sharpe, A. G. (2018). *Inorganic Chemistry* (5th edn), Pearson.

3. Huheey, J. E., Keiter, E. A., Keiter, R. L. (1993). *Inorganic Chemistry: Principles of Structure and Reactivity* (4th edn), Harper and Row.

4. Cotton, F. A., Wilkinson, G., Gaus, P. (1995). *Basic Inorganic Chemistry* (3rd edn), Wiley.

5. Cotton, F. A., Wilkinson, G., Murillo, C. A., Bochmann, M. (1999). *Advanced Inorganic Chemistry* (6th edn), Wiley.

6. Greenwood, N. N., Earnshaw, A. (1997). *Chemistry of the Elements* (2nd edn), Elsevier.

Texts dealing with structural inorganic chemistry and bonding

1. Wells, A. F. (1984). *Structural Inorganic Chemistry* (5th edn), Oxford University Press.

2. Gillespie, R. J., Hargittai, I. (1991). *The VSEPR Model of Molecular Geometry*, Allyn and Bacon.

3. Müller, U. (1993). *Inorganic Structural Chemistry*, Wiley.

4. Alcock, N. W. (1990). *Bonding and Structure*, Ellis Horwood.

5. Hoffmann, R. (1988). *Solid and Surfaces: A Chemists's View of Bonding in Extended Structures*, VCH.

Texts dealing specifically with main group chemistry

1. Massey, A. G. (2000). *Main Group Chemistry* (2nd edn), Wiley.

2. Henderson, W. (2000). *Main Group Chemistry*, Royal Society of Chemistry.

Texts dealing with the general understanding of inorganic chemistry and related topics

1. Owen, S. M., Brooker, A. T. (1991). *A Guide to Modern Inorganic Chemistry*, Longman.

2. Barrett, J. (1991). *Understanding Inorganic Chemistry*, Ellis Horwood.

3. Murphy, B., Murphy, C., Hathaway, B. J. (1998). *Basic Principles of Inorganic Chemistry: Making the Connections*, Royal Society of Chemistry.

4. Winter, M. J. (2016). *Chemical Bonding*, Oxford University Press.

5. Keeler, J., Wothers, P. (2008). *Chemical Structure and Reactivity: An Integrated Approach*, Oxford University Press.

6. Keeler, J., Wothers, P. (2003). *Why Chemical Reactions Happen*, Oxford University Press.

7. Albright, T. A., Burdett, J. K., Whangbo, M. H. (1985). *Orbital Interactions in Chemistry*, Wiley.

8. Gimarc, B. M. (1979). *Molecular Structure and Bonding*, Academic Press.

9. Dasent, W. E. (1965). *Nonexistent Compounds*, Edward Arnold.

Texts dealing with periodicity

1. Barrett, J. (2002). *Atomic Structure and Periodicity*, Royal Society of Chemistry.

2. Puddephatt, R. J., Monaghan, P. K. (1986). *The Periodic Table of the Elements* (2nd edn), Oxford University Press.

3. Mingos, D. M. P. (1998). *Essential Trends in Inorganic Chemistry*, Oxford University Press.

Books containing useful data

1. Emsley, J. (1989). *The Elements*, Oxford University Press.

Index